TRACKS IN THE SAND

TRACKS IN THE SAND

Sea Turtles and Their Protectors

Frank Gromling

FRANK GROMLING
AND MIKE CAVALIERE

OCEAN PUBLISHING

Ocean Publishing
P.O. Box 1080
200 S. Oceanshore Blvd., Suite 4
Flagler Beach, FL 32136
(386) 517-1600 • (888) 71-OCEAN
www.oceanpublishing.org

Library of Congress Cataloging-in-Publication Data

Gromling, Frank & Mike Cavaliere
Tracks in the Sand: sea turtles and their protectors / Frank Gromling & Mike Cavaliere
p. cm.
Includes biographical references and index.
Summary: "Details the plight of the endangered sea turtle as well as the tireless efforts of one Florida-based group who fights to conserve them, the Volusia/Flagler Turtle Patrol." —Provided by publisher.
ISBN 978-0-9826940-0-8 (alk paper)
1. Sea turtles. 2. Endangered species. 3. Turtle Patrol. I. Title.
QC981.8.G56S78 2010
XXX.XX'XX—dc22

2010XXXXXX

Cover Design: Carol Tornatore
Interior Design/Typesetting: Gary A. Rosenberg
Copyeditor: Jane Pugh
Author Portraits created by: Susie Kontul
Interviews transcribed by: Jenni Parks
Cover Photo provided by: Lori Ottlein
Interior Photos provided by: Beth Libert, Lori Ottlein, Dale Hench, Teresa Ryan, Paul Van Lengen, National Marine Fisheries Service, Caribbean Conservation Corporation, Turtle Conservation Centre, Marine Science Center of Ponce Inlet, Mote Marine Laboratory and Aquarium, The Turtle Hospital
Research: Jill Gorey, Stephanie McClain, Lori Ottlein

Printed in the United States of America

10 9 8 7 6 5 4 3 2 1

Contents

"I think it was fate that the turtles chose me to be their protector and spokesperson. I have always been active, rescuing sick and needy wildlife as well as stray cats. It would not have been my choice to be so involved in sea turtle conservation if I had known how it would take over my whole life."

—BETH LIBERT, PRESIDENT OF THE VOLUSIA/FLAGLER TURTLE PATROL

The Beginning

This is how it started. The ocean breeze carried in bits of the evening sun and traced it through Beth Libert's hair. She was on a drive with her boyfriend, David, through the glow of a balmy summer sunset on the coast of Florida. It was 1983. The sky was blushing in shades of pink and gold.

She knew this area well, this blur of roads and beaches just north of the small fishing town of Ponce Inlet. She grew up here. But there was one thing about the familiar streak of shadow, sand and home outside her window that somehow never registered. And then she saw them, a woman and a little girl, digging in the dunes.

"Stop the truck!" she blurted to David. "Those people are digging a sea turtle nest!" It was as though a switch had been flipped within her, the stereotypical light bulb of so many cartoons and clichés coming to life over her head. All of the years that she had spent on and around this beach, splashing through the surf or admiring it from a distance, it never dawned on her, not until this very moment: she was not alone here. Sea turtles nested here. David stopped the truck and Beth got out.

"What are you doing?" she asked the couple, watching them scoop sand from a hole and methodically drop it behind them. It was a seemingly simple question, the kind you ask in passing. The kind you'd never expect to change your life forever.

The woman glanced up and wiped moisture from her forehead.

Her name was Joyce Durrbeck and, with her husband Richard, she held a state permit which allowed them to dig nests in search of turtle eggs to be relocated to their home garage for incubation. Without hesitation, Beth asked if she could help.

Beth wasn't the first to offer her assistance to Joyce or her husband this way. But digging turtle nests wasn't easy work, and so Joyce told Beth to speak with Richard, who was nothing short of pessimistic. "Everybody says they want to help," he told her, "but not many are willing to do the work." After all, he explained, the bulk of what they do is done at night, usually very late. But Beth persisted and he agreed to call her the next time a nest was found that needed excavating.

It was a few nights later at 2:00 A.M. when Beth's phone rang. It was Richard. He had a turtle nesting on the beach in Ponce Inlet and asked if she wanted to help move the eggs.

This was a test. To call this late, for Beth to crawl out from the sound embrace of her bed covers, drive to the shore and plunge herself into the sand, well, she'd have to really want it. This was Richard's version of boot camp. This was him offering the experience of a rude awakening. If Beth wasn't going to last, he probably figured, he might as well weed her out early. This was him opening the door for her to change her mind—no hard feelings. But Beth was thrilled.

Before even hanging up the phone, her feet were on the floor, her car keys wind chimes dangling from her fingers. She obviously said yes—she even arrived at the beach before Richard did—and together they dug. One by one, they placed their eggs inside the Styrofoam cooler he brought along to every job. And one by one, they bonded.

In 1983, the state required that all sea turtle nests in that part of Volusia County from northern Ormond-by-the-Sea to southern Ponce Inlet be emptied and moved to a safe offsite location at which to incubate. At this time, driving was still allowed on all of the beaches in the area, and turtle nests were routinely run over and destroyed, hatchlings squashed before they could even take their first steps into the ocean. The Durrbecks had set up in their garage a veritable nursery for turtle eggs, sometimes with more than fifty coolers stacked in corners and along the walls. These coolers, these Styrofoam wombs, were all full of eggs, every one laid by mothers who would never return to see their babies hatch or help them through the water.

In this way, the Durrbecks' garage was transformed into an orphanage, a cabin crowded white with dozens of coolers resting on top of one another like bunk beds. From then on, whenever Richard and Joyce would turn down the lights and head inside after each new dig, it was as if they were saying goodnight to a hundred sleeping infants, gently turning the knob as they closed the door so as not disturb them.

Eventually Beth, too, turned her garage into an incubation center, and soon her name was added onto the Durrbecks' license, granting her the authority to work independently and essentially doubling their collective response along the beach. The three had officially formed a team, the stems of a root structure that would later give bloom to a canopy that would extend across cities, hanging like a ceiling of eyes and ears above beaches and tides, watching over them, protecting them. They became the prologue to a story whose end they'd never imagined—a patrol of hundreds, a TURTLE PATROL.

Note: All Patrollers showcased throughout these pages were chosen at random to reflect the highest diversity of perspectives.

CHAPTER 1

Let's Talk Turtles

"OH just wait!!! When you see the mama tracks in the sand and then the nest!! WHOO!! Around July when the babies hatch—*then* the OMG comes on!!"

—TEN-YEAR TURTLE PATROL MEMBER LORI OTTLEIN, EXCERPT FROM E-MAIL

Prehistoric in appearance and with a navigational savvy that would humble Magellan, sea turtles inhabit temperate and tropical oceans throughout the world. These magnificent sea-dwelling reptiles can migrate hundreds, sometimes thousands of miles returning to nest on or near the very beaches from which they first emerged as hatchlings. An ancient species existing more than two hundred million years, sea turtles once swam with dinosaurs and numbered in the millions. Over the last few hundred years, however, their populations have plummeted, and now, of the seven species of sea turtle on earth—six of which can be found in U.S. waters—every one is listed under the Endangered Species Act.

Today, groups of activists all over the globe work tirelessly to save these creatures. These stewards are generally volunteers, organizing patrols and monitoring nesting beaches in their free time, pro bono. They collect data and inform the public about the hazards of nighttime beach activities, beachfront lighting, human intervention, and pollution. One such group working diligently on behalf of these causes is the Volusia/Flagler Turtle Patrol.

Officially founded in 1984, the Turtle Patrol has existed in some way, shape or form in communities forever. A small yet passionate bunch, they're the few you see on the shore just before sunrise, watching night fade away as the sounds of their city waking up grow louder and louder around them. They're the ones with sandy knees and ATVs (all-terrain vehicles), wooden stakes and tee-shirts, sectioning off parts of the beach where life might later blossom. They're home there, at the edge of the world, standing barefoot in the sand and squinting deep toward the horizon, wondering just how many of the babies they helped birth are out there, swimming, diving and living out there, warm underneath the rippled sheet of ocean tucked just beneath the sun.

* * *

Today, Florida's Volusia/Flagler Turtle Patrol is a group of a couple hundred volunteers led by Beth Libert who monitor fifty-nine miles of beach along Florida's east coast where sea turtles nest. They conduct these surveys in the dead of summer, traversing their given stretches of coast in shifts, protecting nests and relocating those laid in less-than-optimal emergence conditions. Seeing every nest through until egg hatching, Patrol members volunteer their time year-round, recording birthing numbers, seeing infant turtles into the ocean, and educating the surrounding population about sea turtles and their plight. Every Patroller, no matter where they're initially from or how they came to first enlist, has one major and unifying trait in common: they all share a striking, insatiable, and sometimes difficult to fully understand passion for their work. They "love" turtles, a love most of them can't clearly articulate. They don't call it spiritual, their bond with the beach. You'd never hear them refer to the sand as a confessional or the sky stained glass. They only smile, use words like "duty," "fulfilling." They only laugh, and talk about the first time they ever saw a nest emerge, and how from that moment on, they were addicted.

WHAT'S THE DEAL WITH SEA TURTLES?: THE BASICS

Sea turtle species vary in shape, size and color but all have streamlined bodies perfectly adapted to life in their aquatic environment. Using long, paddle-like flippers, they appear calm, graceful as they glide through the water as quickly as thirty-five miles per hour.

Although on land a turtle's vision is weak and short-sighted, underwater it is sharp, as is its sense of smell. Their acute sense receptors extend even to their shells. Despite its rugged appearance, a sea turtle's shell is actually sensitive to touch, the same as are parts of its flippers. Lacking external ears, turtles rely upon a single bone in the middle canal to conduct vibrations to the inner drum, giving them the ability to respond to low frequency sounds.

Sea turtles live in coastal waters, bays, lagoons, estuaries and the open sea. Air-breathing reptiles, they remain submerged for long periods of time, surfacing to catch quick, but full, breaths of air. Each rapid inhalation replaces a much higher percentage of oxygen in its lungs than that of a comparable human breath. Some species, such as greens, can even rest or sleep underwater for several hours at a time as their heart rate slows to conserve sustainable air. In these circumstances, a whole nine minutes might elapse between the heartbeats of a green sea turtle. Other species, such as leatherbacks, store large quantities of oxygen in their muscles and blood.

Although sea turtles are wonderfully adapted to life in the waves, it remains entirely possible for them to drown. When entangled underwater, for instance, by fishing line or net, stored oxygen is quickly exhausted in frantic efforts to free themselves.

Like all reptiles, sea turtles are cold-blooded. Their body temperature reflects that of their environment.

Happiness is a Warm Shell

Note: While putting this book together during an unusually cold Florida January, reports began flooding in of mass cold shock occurring among turtles all along the coast, primarily greens. It happened that the temperature dropped so far below what the turtles were used to or able to withstand that it led them to go into a state similar to human hypothermia. Their bodily functions shut down, and their muscles atrophied to a point where they were unable to raise their necks above the surface to breathe (the same condition that causes iguanas to fall out of trees when it gets too cold). Soon thousands of shocked or dead turtles began washing up onto the beach. The Turtle Patrol did what they could to revive and transport as many casualties as possible to local marine centers, which were so overwhelmed by the influx that employees took to working twelve-hour days and housing/treating these animals in dozens of kiddie pools lined down their hallways and inside their offices.[1] In some cases, simply getting the turtles out of the water and onto sun-drenched pavement would begin bringing them back to life; in others, it would take more intensive rehabilitation.

In any case, without the efforts of these individuals, many if not all of the cold-stunned turtles would have been lost to the weather, a huge detriment to an already trying species. In just three days in January, though, hundreds were admitted and nearly 200 saved by The Turtle Hospital in Marathon, Florida, which is more than the average of animals the center usually admits in an entire year. Many of the more than 1,000 washed up still didn't make it, but because of those who helped, hundreds survived to see February.

From the Top: The Shell

A sea turtle's shell is large and bony and provides protection from injury and predation. The dorsal, or top side, is called the carapace, and the ventral, or bottom side, is called the plastron. Covered with a layer of firm but flexible plates called *scutes*, every species of sea turtle with the exception of the leatherback has a heavy, hard-armored shell that protects its torso from predators. The leatherback's shell differs in that it has a thick, oily skin covering the plates, making it a softer, more pliable texture than the others. Scientists can identify the various species of turtle by examining the number and pattern of scutes on its shell. This is especially helpful when differentiating between turtles of similar size and color, such as greens and loggerheads.

Of the several defining characteristics separating sea turtles from their land-dwelling relative, the tortoise (one is home in water and one will drown if submerged in a mere three inches of it; one eats fish and seaweed, the other eats fruit and lettuce; etc.), the differences in shell seem to be the most readily observed of the bunch. Unlike the tortoise's curved and dome-shaped carapace, the sea turtle's shell is flatter and more streamlined. It also serves different defensive purposes. Unable to retract its head or limbs inside of it, the sea turtle's shell is more for padding and aesthetic protection than full body armor or retreat.

If the Sand Is A-Rockin'. . . : Nesting Behavior

For a female sea turtle, the journey from sea, to land, to nesting spot is a vulnerable one. Their heavy, cumbersome bodies and flipper-like limbs were not designed for movement on land. Males rarely, if ever, return from the water, but females must come ashore to lay their eggs, fighting the elements

and unforgiving terrain up the beach to birth their clutch. Beaches with deep, loose sand are optimal as egg embryos breathe outside air through their shell membranes. For this reason, turtle eggs cannot survive if immersed in water. In the United States, nesting seasons occur from April through October.

Leatherbacks are usually among the first to appear on Florida beaches, given their ability to withstand cooler temperatures. Other varieties, as a rule, nest later in the season. No matter when first nesting occurs, however, every species performs the process a different number of times per year. Some might nest only twice or a single time after mating; others can drop as many as ten clutches.

As far as courtship and mating is concerned, the sea turtle version of the age-old "Birds and the Bees" story isn't so much a silky Barry White-type romance as it is a thrashing Van Halen one. Open to mating only during a short "receptive" period, males may nuzzle the head of a desired female or softly gnaw on her neck and rear flippers. If the female doesn't high-tail it after this advance, the male will climb onto her back, grip her top shell with his claws and front flippers, and fold his tail under her shell to copulate. After that, it's a simple waiting game before the female returns to land to birth her clutch, where surveyors have often observed scratched shells and bleeding body cuts from the claws of past mates.

Mating among turtles is an aggressive act, often with several males competing for one female, and one female mating with a handful of different males before nesting. This medley of sexual partners, scientists believe, may play a crucial role in maintaining a high genetic diversity among the species.

Generally coming ashore to nest at night, alone and during high tide, females crawl beyond the surf, find an ideal location and begin digging a cavity in the sand. They will labor at this task for over an hour, flinging away dry sand with their flippers, rotating their bodies in circles to construct a "body pit" in the divot, then shoveling a final tear-shaped egg chamber into the ground before depositing fifty to one hundred fifty pliable, rubbery, Ping Pong ball-shaped eggs into the nest. Using their hind flippers, they fill in the chamber and fling sand with their front flippers over the pit to camouflage the area before making the journey back into the sea.

Sea turtle egg chamber
A sculpture by Tom McFarland

Bright-Eyed and Bushy-Tailed: Hatchlings

After an average two-month incubation period, eggs begin to hatch. Hatchlings break through their shell using a carbuncle or temporary egg tooth, and, with nests generally dug two feet into the sand, it can take hours, sometimes days, for them to make it to the surface. This first crawl, or emergence, usually occurs at night and in groups. The first to emerge serves as a kind of trigger to all the rest, using the pyramid of sand and breaking egg shells inside to crawl its way onto the beach in a growing, collective mass. Weighing less than an ounce and with eyesight barely extending past the sand before them, hatchlings use light cues to make their way towards the natural glow of the open horizon. Some will make it to the water; many will not. Hatchlings are easy prey for birds, crabs, raccoons, dogs, and even ants. Their movement can be impeded by seemingly harmless things: tire tracks, holes in the sand, objects on the beach, even a footprint—all small things to man, but giant obstacles to tiny, two-inch long creatures. Hatchlings are also led astray by artificial street and residential light which they mistakenly follow away from the water.

'Hello!! Coming out NOW!!! Just dug myself out of two feet of sand and I am ready!!'

Caption by ten-year TP member Lori Ottlein

Especially interesting in the nesting/emergence process is the enormous role that temperature plays in egg incubation. To a large degree, sand temperature—that is, *nest* temperature—is the factor that determines the sex of a hatchling. Warmer temperatures tend to produce more females, whereas cooler temperatures produce more males. So crucial is temperature to hatchling development

I'M NOT CRYING: THE FALLACY OF TURTLE TEARS

While consumed in what many refer to as a nesting "trance-like" state, a sea turtle may appear to cry. What it's actually doing is ridding its body of excess salt. Sea turtles have large tear glands that concentrate the brackish compound then empty it out through their eye sockets, a process the more sentimental observer might construe as tears shed for a clutch of babies a mother will never meet. This liquid flowing from the eyes also helps to keep them free of sand as a mother digs her nest.

that something as seemingly inconsequential as a rainstorm lasting several days can cool the sand enough to completely alter a beach's gender percentages. Extended periods of unseasonal weather and even the amount of shade over a nesting area can determine the gender ratio of a mother's clutch.

Growing Up Turtle: The Awkward Middle Years

On making it through the crashing waves and the thrashing tide of the beach, hatchlings swim for miles until they are caught in offshore currents and seaweed, flotsam and jetsam, which chauffeur them further out to deep-water convergence zones. There, in these critical first forty-eight hours, they begin their "drifting cycle," floating passively in mats of seaweed and algae that are home to jellies, shrimp, snails, crabs, and many other marine delicacies on which they'll feast. This cycle will continue usually for several years, until post-hatchlings grow to about dinner-plate size, then eventually move into coastal, neritic zone (shallow near-shore waters with muddy or sandy bottoms on which prey can be easily found) waters to continue their maturation. The hypothesized length of the Kemp's ridley drifting cycle, for example, is roughly two years; seven to twelve for the loggerhead. In the meantime, post-hatchlings just float, eat, and grow.

In addition to providing an essential habitat for juveniles, the neritic zone is also a prime foraging grounds, perfect for inter-nesting, and even a migratory habitat for adults. Although many of these different ocean zones may overlap, serving several different uses depending on a turtle's particular stage in its life cycle, make no mistake, the sequence in which these zones are inhabited is critical, especially for post-hatchlings. With their undeveloped bodies and utter inability to ward off predators, post-hatchlings depend on the isolation and abundance of food sources found in convergence zones in the same way that juveniles depend upon corresponding circumstances in neritic areas. As well as changing their living quarters, turtles moving from one ocean zone to the other also change their eating habits. Diets that used to consist primarily of floating surface grub seamlessly evolve into a higher dependence on sources below the surface, organisms found in coral reefs and similar underwater locales. To continue to grow and mature, turtles must move to more shallow sea regions capable of supporting their development. Only in these new ocean zones can they find the appropriate diet and habitat.

Commitment Issues: The Migration Shuffle

Migration patterns of most species of sea turtle can hardly be defined, let alone broken down into logical, linear graphs. Although females of certain species are known to follow set paths from mating to nesting grounds, with foraging zones serving as sort of rest-stops along that pre-determined map, they don't necessarily follow this same route every year, or nest every year at all, for that matter. Many male turtles may never migrate—and if they do, tracking their routes can prove an odd and dumbfounding process. It is the olive ridley, in fact, that scientists label as "wandering" migratory creatures, because of their indefinable routes and abrupt changes in direction.

Madam X, tag # X9244, has had two of her nests, #N9 and #S22, recorded by her tracking harness.

Couple the seeming inconsistencies of course with the fact that for much of the time that turtles are at sea they remain deep underwater, even sometimes sleeping underwater, tucked below a rock or reef, and not even decades of research have been able to iron out the logistics behind their enigmatic journeys. Migratory routes, in this respect, just might be the key to uncovering difficult-to-pinpoint hotspots requisite to the species' survival. Once these habitats have been identified, scientists and conservationists would know better where to focus their efforts. The simplest and most inexpensive tracking method used today is the employment of small, harmless metal tags placed on a turtle's flipper. On smaller turtles, tiny microchip tags, known as pit tags, are imbedded in the shoulder. These are basically the equivalent of a turtle collar, scripted with a coded number and a message asking people to return the tag to an included address if found. On return, the person who found the tag is usually given a small reward and asked where the tag was found. Through this process, over time, a map is formed. Given enough maps, patterns can be established.

Another higher-tech alternative is the use of satellite telemetry. With a Walkman-sized transmitter attached to a turtle's back, researchers can track movement every time a turtle-subject breaches the surface to breathe, lifting the transmitter's antenna from the water in the process. After eight to ten months, this transmitter will no longer work and eventually fall from the shell. By then, researchers will have procured a solid amount of data to be analyzed.

Despite the all-over-the-place, meander-like approach of some sea turtles, the migratory routes of others—namely the nesting females of each species—were found to be fairly consistent according to data recorded via satellite or tag. Take the Pacific loggerhead, for example, and its straight-line approach from nesting beaches in Japan to feeding grounds off Mexico. The other extreme is the olive ridley, with its seemingly unplanned paths and patterns from one foraging zone to the next. Although many of these paths might, to us, seem perverted and often baffling, one has to wonder if there remains some kind of inherent, instinctual method behind the apparent madness. For now, motivations remain unclear, leaving the worlds of science and human ego with the humbling reality that no matter how advanced or evolved we believe ourselves to be, no matter how many satellites we propel into orbit or how many equations we design to map out our place in the world, the behaviors of one of the oldest species on the planet can still astound us.

Dale and Peach Hench

Members of the Volusia/Flagler Turtle Patrol:
2003–Present

"It's just fun. That's one reason we do it, because it's fun"

Talk to anyone in the Turtle Patrol and they'll speak of Dale and Peach Hench with an uncanny sense of respect. If Beth Libert is the elected "mother" of the group, Dale is certainly the grandfather. With salty white hair, a vibrant tenor, a coarse, full beard and a background in environmental work with Illinois state parks, it makes sense that he's looked on with such repute, and why he came to occupy the position of official presentation conductor at nest cleanings and schools.

"Dale is phenomenal at standing there and talking to a large group of people," seven-year Patroller Barbara Ryan said. "People just gravitate toward him," her daughter Catherine agrees. You look at him, with the beard and the voice, they reflected, and it's hard not to think, "This guy's *got* to know what he's talking about!

Whereas other members interviewed tended to veer their conversations more toward the personal elements of patrolling, their emotions and experiences, the Henches stayed closer to the day-to-day aspects of how surveys are conducted and why. Their passion for the work is evident—it spills out and over their every word, even bleeds into their fashions. Dale wore a Marineland Right Whale Project tee-shirt and a necklace with a large sea turtle charm hanging over it, resting against his chest. They talked with us for nearly two hours.

Dale presenting to a group of kids
on the beach before a nest evaluation.

[excerpt from interview]

Why do you do this? Why do you get involved?

Dale: Just our love of being outdoors. We have the advantage of being retired so we can devote time to that sort of thing . . .

So it's more about the outdoors and nature than the animals?

Dale: No . . . it's addicted to the animals too, yeah—and giving back. We're involved with human activities and charities as well, but this is just something that appealed to us. The animals are threatened and/or endangered and they need our help. [We've been involved with other conservation initiatives] but this was just a little bit more hands-on and was *ours,* meaning the Turtle Patrol and people who had a like mind.

Peach: It's a privilege to do it. With the whales and turtles, you know, there's so few of them in many respects. And [people will say], 'What about the turtles?' 'Why are you doing all this work?' 'They're just good for eating,' or something like that. And I say, You know what? They've been around for a hundred million years and I don't want them to go away while *I'm* still alive! They've been around all these years.

Dale: Plus it's a learning experience. You can read about something in a book or watch it on the nature channel, but then to have a hands-on experience, to hold a little sea turtle in your hand, assist in the recovery or a release after its been to the turtle hospital . . . you know, you can't do that except by doing it.

Sounds like a lot of it has to do with Florida, being here.

Peach: It is. I mean, we love it here, every day.

Dale: And we're not unique. Our love of it and enthusiasm is matched by everybody else's.

Peach: [You have to have that.] You wouldn't be there if you didn't.

[Nearing the tail-end of their interview, Dale and Peach reflected on the personal rewards of their work.]

Dale: We've had kids that you would think might be jaded. You know, they like their video games and they're on the beach because their folks made them get on the beach. They're the ones we've had people send us, 'This is my son's report on What I Did During My Vacation.'

Peach: Oh yeah, a lot of them. 'This is what we're gonna write about when we go back to school.'

That's a big reward.

Dale: Yeah, it really is. It's not us, specifically. It's just the whole process, the fact that . . .

You're a part of it.

Dale: Yeah. . . . You can't beat the hands-on experience.

CHAPTER 2

Species Breakdown

Seven species of sea turtle exist worldwide, and six of them can be found in U.S. waters. Of these six, five species nest along the beaches of Florida: the loggerhead, green, leatherback, hawksbill and Kemp's ridley. All six sea turtles found in the United States are either of "Endangered" (green, hawksbill, Kemp's ridley, leatherback), or "Threatened" (loggerhead, olive ridley) status and are protected under the Endangered Species Act of 1973.

THE BIG SIX

LOGGERHEAD *(CARETTA CARETTA)*

The most common sea turtle to nest in the southeastern United States, the loggerhead was named for its relatively gargantuan cranium which supports powerful jaws that allow it to feed on whelks, conk, lobster, shrimp and other hard-shelled prey. Its shell is a dense, flaky thing which can become spongy with all manner of growth as many smaller species of plants and animals attach themselves onto it. One loggerhead found in the inlet of Ocean City, Maryland, could no longer dive due to a whopping ten pounds of attached organisms living on its back, including barnacles, blue mussels, and sea grasses.[2] If not for a fortuitous rescue and rehabilitation, this turtle would have starved or been in great peril of injury or death by boat collision.

From hatchling to adult, a loggerhead increases its weight more than six thousand times. With average carapace lengths of approximately thirty-six inches and average weights of 250 pounds, they reach sexual maturity anywhere between twenty to thirty-five years of age. At this time, mating begins, usually occurring in late March to early June, with females laying eggs between late April and early September. Incubating for roughly two months, these eggs will hatch sometime late June to mid-November. Females will lay three to five nests, even sometimes more, containing an average of one hundred eggs per nest, during a single nesting season.

During the several months that a female loggerhead breeds, she will travel hundreds of miles to lay thirty-five or more pounds of eggs before swimming all the way back home to her foraging area, all without eating anything substantial. This ability is not unique to Atlantic loggerheads, though. Pacific loggerheads are known to travel over 7,500 miles between nesting beaches in Japan and feeding grounds off the coast of Mexico. Clearly, sea turtles were built to travel the ocean. A *pelagic,* or free-roaming species, loggerheads are circumglobal, meaning they go where they want, when they want, and occur throughout the temperate and tropical regions of the Atlantic, Pacific and Indian Oceans. They can hold their breath for vast amounts of time underwater and even dive to depths of thousands of feet. It is clear that loggerheads are at home in the ocean, even possessing callus-like traction scales on the bottom of their flippers that make it easy for them to walk, or crawl, the ocean floor for near-shore hard-bottom feeding.

The most abundant turtle species found in U.S. coastal waters, loggerheads prefer relatively narrow, steeply sloped, coarse-grained beaches for nesting. This is why the east coast of Florida (shores along Brevard, Indian River, St. Lucie, Martin, Palm Beach and Broward Counties) is their most important nesting ground in the world. In fact, Florida's Archie Carr National Wildlife Refuge (ACNWR), the twenty-mile stretch of coastline from Melbourne Beach to Wabasso Beach, is the most essential loggerhead nesting site in the entire western hemisphere. In this area alone, researchers have recorded nesting densities of one thousand nests per mile.

When hatchlings emerge, they enter a state of frenzied activity, moving from their nest to the surf, swimming and drifting through the tide, then paddling away from land for what is sometimes several days. After this period, they reside where surface waters converge to form downwellings, entering into a much more passive, lower-energy state. Because of the large amounts of floating ocean debris in downwelling areas—seaweed, etc.—post-hatchlings will simply float and wait to forage on the wide variety of materials drifting past them. They'll linger in these areas for months, allowing currents to carry them within the Gulf of Mexico and the North Atlantic.

Once transported farther offshore, juveniles enter into the oceanic zone where they have been primarily studied in the waters around the Azores and Madeira. Research is conducted elsewhere, but in this region the turtles generally spend most of their time within the top fifteen feet or so of the water column, making their study a far simpler endeavor.

It's not until loggerheads reach adulthood that they begin feasting on hard-shelled prey, crushing shells and bones with their powerful jaws which have earned them the somewhat clumsy nickname "the junkyard dogs among turtles." After reaching this phase of their lives, western North Atlantic loggerheads feed predominately in relatively shallow continental shelf waters of the U.S., Bahamas, Cuba and the Yucatan Peninsula, Mexico. The Florida Bay, which is on the west side of the Florida Keys, is also an important feeding ground, most likely because of its easy access to open ocean and migratory routes.

Because loggerheads feed on floating prey and are not typically bottom foragers, they are more often lured to baited hooks fishermen set floating in the water than are other turtle species. For this reason, the primary threat to their populations worldwide remain incidental capture in commercial fishing gear, primarily longlines and gillnets, but also trawls, traps, pots and dredges. In Florida, their number one cause of death is collision with boat propellers. Nesting and harvesting-related losses are also relevant concerns, but the loggerhead's greatest cause of decline—that is, the greatest threat to its survival—remains encounters with these different fishing apparatuses. This industry is exactly what conservation efforts have been focused so fiercely on regulating for years, rallying to establish stricter fishing laws, or, in the least, the use of larger circle hooks and turtle excluder devices (TEDs) in fishing nets.

The loggerhead turtle was listed under the Endangered Species Act (ESA) as threatened throughout its range on July 28, 1978.

GREEN (*CHELONIA MYDAS*)

The only sea turtle to exclusively eat plants, the green turtle gets it name from the earthy color of its body fat. It's the largest of all hard-shelled turtles and is a strict herbivore, eating only algae and sea grasses even through adulthood—which it grazes on the same as cattle or manatees, cutting an entire patch down to equal length before moving on to another. Keeper of a flattened, oval body, adults can grow to larger than three feet long and weigh anywhere from 300 to 350 pounds. Their carapace varies in color from anything in the green family to bright yellow with brown and green striping.

With a thirty-year margin of error in place, scientists haven't exactly determined when green turtles reach sexual maturity—it can be anywhere between twenty and fifty years of age, they say. Their nesting habits, however, have by now been fairly well documented. Females return to their natal beaches (the same from which they were born) every two to four years to lay eggs, nesting at approximated two-week intervals and laying an average of five clutches a season. In Florida, green turtle nests usually incubate for a solid two months and yield almost 135 eggs. Depending on location, nesting seasons will vary. In the southeastern U.S., nesting usually occurs between June and September, with a peak in June and July.

Green turtles are distributed around the globe, thought to inhabit more than 140 different countries' coastal regions worldwide, nesting in eighty or so of those throughout the year. Two of their most important nesting beaches, however, are found at Tortoguero on the Caribbean coast of Costa Rica, and Raine Island on the Great Barrier Reef in Australia, where a yearly mean of 22,500 and 18,000 females nest per season, respectively. In the United States, the central and southeastern coasts of Florida are hotspots, with numbers estimated anywhere from 200 to 1,100 females annually. Important feeding grounds in Florida include the Indian River Lagoon, the Florida Keys, Florida Bay, Homosassa, the Crystal River and Cedar Key.

Despite the apparent stability of the green turtle's nesting numbers, in actuality they have been in a state of steady decline for some time. Studies conducted by the Marine Turtle Specialist Group (MTSG) have shown substantial population dips occurring in *all* major ocean basins over the past 100–150 years. Furthermore, the population trends at thirty-two index nesting sites worldwide showed a forty-eight to sixty-five percent relapse in the number of mature females nesting annually over the same timeframe. Because of these losses, the MTSG of the World Conservation Union (IUCN) officially classified green turtles as endangered globally in 2004.

The main offender to this specie's global decline is the long-term harvesting of eggs and adults from nesting beaches and juveniles and adults from feeding grounds. Similar to other species of sea turtle, these harvests are carried out for purposes of consumption, sale on the black market, or for use in the production of various products utilizing turtles' oil and shell. Even though many federal and international laws forbid this kind of trade—including sections

Fibropapillomatosis: FP for Short

First discovered in 1938, fibropapillomatosis is a recent epidemic in the sea turtle community. A skin disease marked by an outbreak of benign but debilitating cutaneous sores, FP is thought to be tied to a herpes-type virus. These tumors will multiply and can eventually obstruct a turtle's movement, eyesight and breathing passages. Although it is possible for these symptoms to spontaneously regress, surgical extraction appears to be the only proven cure.

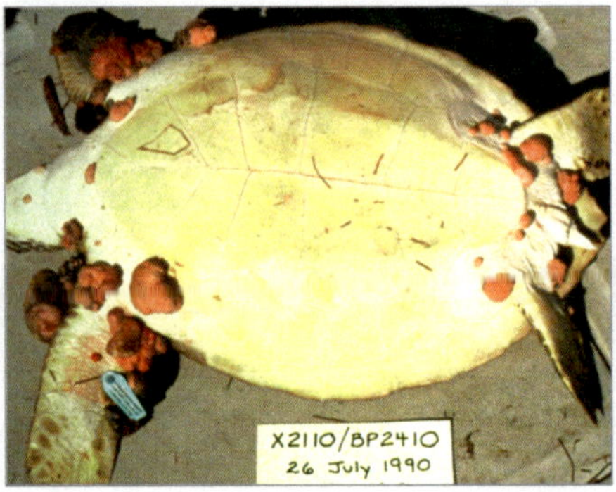

The decline of the green turtle population can be felt throughout coastal ecosystems everywhere. Sea grass beds, for example, must be constantly cut short and maintained to promote healthy growth and expansion. In this, sea grass is just like normal lawn grass. If these beds are not maintained, blades will grow longer but will not expand to other areas of the sea floor, containing fewer and fewer nutrients the longer they're left uncultivated. Alarmingly, the past few decades have shown a harsh decline in ocean grasses, which many scientists trace back to the overall decline of turtles in the sea. The chain is simple: the fewer sea turtles there are in the water, the fewer sea grass beds will thrive, leading to terrible losses of important breeding and developmental grounds for all kinds of fish, crustaceans, and shellfish. If sea grass beds are lost, it could eventually result in the loss of every species that depend on them, as well.

The green turtle was listed under the ESA on July 28, 1978. The breeding populations in Florida and the Pacific coast of Mexico are listed as endangered; elsewhere the species is listed as threatened. Critical habitat was designated in 1998 in waters around Culebra Island, Puerto Rico.*

*Critical habitats are, simply, specific regions deemed pivotal to a threatened species' survival, based on unique resources found there. For the North Atlantic Right whale, for example, the southeast tip of Georgia and the northeast part of Florida are "critical" because it is in those regions where the whales birth their calves in the warmer southern waters.

of the Convention on International Trade in Endangered Species of Wild Flora and Fauna (CITES) and the Convention on Migratory Species (CMS)—these harvests continue in certain parts of the world, undermining the efforts of activist groups and conservationists alike. Similar to loggerheads, capture in fishing gear is also a serious source of mortality. In some regions of the world, a skin disease known as *fibropapillomatosis* (FP) is taking its toll, as well (see the inset on the facing page).

LEATHERBACK *(DERMOCHELYS CORIACEA)*

The leatherback is a god among sea turtles. It's the largest turtle in the world (as well as the largest living reptile); it travels the farthest (they've been found more than three thousand miles from their nesting beaches); and it dives the deepest (sometimes deeper than 3,900 feet). But wait, there's more: with an average adult length of nearly six-and-a-half feet, the largest leatherback on record was a male stranded on the coast of Wales in 1988 that weighed almost 2,020 pounds. If there is any species that can make you truly appreciate how closely the turtle's bond is to the dinosaurs, it's the leatherback. It's a bona fide contradiction in time, a souvenir of something lost and scarcely seen again.

The only soft-shelled sea turtle in the world, the leatherback's carapace is composed largely of cartilage, oil-saturated connective tissue overlaying a series of interlocking outward-growing dermal bones. This Volkswagen Beetle-sized creature derives its name from the leather-like skin that covers and protects the bony plates of its shell. This skin, as well as thermoregulatory adaptations, a counter-current heat exchange system, high oil content, and large body size, serve as excellent insulation allowing the leatherback to tolerate much colder waters than that of other species of ocean-dwelling turtle. This higher tolerance also makes leatherbacks the most migratory and wide-ranging sea turtle in the ocean, with specimens spotted as far off the eastern U.S. as Newfoundland. Other subjects, tagged and tracked by satellite transmitter, have been recorded off Nova Scotia and followed to waters adjacent to nesting beaches along the northeast coast of South America, the Antilles, Panama and Costa Rica.

Leatherbacks mate in waters near nesting beaches and along migration paths. On leaving their carefully laid eggs to incubate, females migrate from tropical waters to more temperate ones, which support higher densities of jellyfish during the summer months. Because leatherbacks lack the powerhouse crushing jaws typical of other sea turtles, these jellies prove an essential food source. Instead of feeding on hard-shelled prey, leatherbacks are adapted to consume and keep down gelatinous creatures. Their pointed tooth-like cusps, sharp-edged jaws and backward-facing spines are perfectly designed for an invertebrate-heavy diet. In this regard, they're not unlike the North Atlantic right whale, another mammoth-sized marine animal who feeds on tiny, non-threatening prey. It's interesting that creatures of this girth and magnitude, creatures that might otherwise frontline the food chain, gather their food rather than stalk it, whether it's jellyfish for leatherbacks or tiny copepods* for right whales. It's an odd incongruity when compared to the diets of other oceanic beasts, like Great White sharks or orcas. Despite great size, in nature, bigger doesn't always mean more ferocious. Turtles, the same as countless other giants, are simply not hunters.

After a leatherback nest hatches, the research on juvenile distribution and developmental habitats gets fuzzy. Although scientists know where the largest remaining leatherback nesting assemblages are located around the globe—on the coasts of northern South America and west Africa—exactly where post-hatchlings go after emergence, where they mature and grow, is poorly understood.

*Copepods are groups of small crustaceans that float in the ocean in a soup of plankton and seaweed. They are found in abundance in certain locales, and serve as important sources of protein.

Perhaps this is why the World Conservation Union has observed more than an eighty percent population decline of this species in the Pacific. True, this decline could also be due in part to the difficulty of monitoring these creatures, being they travel such immense distances and nest on different beaches every year, making it nearly impossible to keep a steady eye on rising or falling trends. But if more were known about the location of post-emergence zones, more could be done to ensure that those areas were protected and that statistics were legitimate. The same holds true for all free-roaming aquatic species.

That being said, the known primary threats facing leatherbacks today, as they are for loggerheads and greens, are long-term harvesting and incidental capture in fishing gear. Capture occurs most often in gillnets, but also in trawls, traps and pots, longlines, and dredges. Harvesting occurs on nesting beaches and feeding grounds where eggs and turtles are most vulnerable.

The leatherback was listed under the ESA as endangered in 1970. Critical habitat was designated in 1998 in coastal waters adjacent to Sandy Point, St Croix in the Virgin Islands, and regulations to require TEDs in shrimp trawl fisheries were enacted in 1992.

HAWKSBILL *(ERETMOCHELYS IMBRICATE)*

The hawksbill is a small but beautifully-marked turtle that got its name from the beak-like way its face comes to a point at the nose, allowing it the ability to reach into reef crevices for food. Growing to weights of 100 to 150 pounds (sometimes as much as 200), the hawksbill isn't exactly tiny, just more compact compared to its relatives. Its carapace can range anywhere from twenty-five to thirty-five inches in length

and most closely resembles that of the tortoise's (dark, sometimes golden brown, with streaks of black, red and even orange). It's this eye-catching shell that brought the hawksbill to its present near-extinction status, due to poaching for use in jewelry, hair decorations and other ornamental items for generations. Like the leatherback, the hawksbill feeds on a variety of invertebrates. Its main diet, though, is composed of toxic sea sponges largely constructed of tiny, glass-like needles which seem to cause the turtle no harm—especially in the Caribbean, where an adult hawksbill will consume an average of 1,200 pounds of these sponges yearly.

Every two to three years female hawksbills return at night to their natal beaches to commence a new nesting season. During this time, they return approximately every fourteen to sixteen days to lay three to five nests, each containing about 130 eggs. These nests are usually dug high on the beach under or in beach/dune vegetation and, even though hawksbills are the fastest nesters of any sea turtle—completing the whole process in less than forty-five minutes—nesting still remains one of the most vulnerable periods for the animal, putting it in great risk of "turtleknapping" for sale and harvest on the black market.

Since its inclusion in the Endangered Species Act, however, poaching has become a less pressing threat to the hawksbill than it used to be. Instead, the greatest danger facing this species today is the continuing degradation of its habitat. Like all turtle species, hawksbills inhabit different locales at different stages of their life cycle—the pelagic (open ocean) environment for post-hatchlings, coastal foraging grounds for juveniles—but healthy coral reefs are where they spend a majority of their adult lives. A credit to the hawksbill's circumtropical nature, these reefs can be found in a great variety of places: the Atlantic, Pacific, and Indian Oceans—especially in the Caribbean and around Florida (on the reefs off Palm Beach, Broward, Miami-Dade, and Monroe Counties), the Gulf of Mexico and the Greater and Lesser Antilles. Really, except for the Mediterranean Sea, hawksbills can be found in any stretch of ocean where healthy coral populate. The problem, of course, is that strong reef systems are becoming more and more of a rarity these days. Exceptionally vulnerable to pollution, human interaction and even climate change, every day reefs are destroyed or otherwise degraded. As a result, the habitat on which hawksbills and more than 530 other marine organisms so rely for food sources as well as habitat are quickly becoming scarce, in both quality and quantity.

In the past, commercial exploitation played the part of the arch nemesis in the hawksbill's fight-away-from-extinction story, and although this threat has been greatly neutralized in recent years, there remains a demand for the hawksbill's shell, as well as its skin and oil for use in leather and cosmetics. In fact, the British Virgin Islands, Cayman Islands, Cuba, Haiti, and the Turks and Caicos Islands (U.K.) all still permit some form of legal take of these turtles. In the northern Caribbean, hawksbills are directly harvested for their carapace. They are harvested for eggs and meat, as well; their bodies are even sometimes stuffed and sold as curios to tourists. They are harvested for products openly available in the Dominican Republic and Jamaica, despite prohibitions. In the Pacific, the trade is all too widespread, with directed take of eggs and nesting females being conducted regularly in American Samoa, Guam, the Republic of Palau, the Commonwealth of the Northern Mariana Islands, the Federated States of Micronesia, and the Republic of the Marshall Islands.

But, again like the leatherback, hawksbills are solitary and sporadic nesters which makes determining their population estimates (read: declines) extremely difficult. Of their behavior, we know that they are known to inhabit the same resting spot night after night, usually on or around reef ledges and caves, and that they can be found around rocky outcrops and high-energy shoals, optimum sites for sponge growth. We also know that their largest nesting population in the U.S. appears in Puerto Rico and the U.S. Virgin Islands, worldwide in Australia—with two thousand hawksbills laying eggs on the northwest coast and about six thousand to eight thousand off the Great Barrier Reef each year. Because of their highly migratory nature, the hawksbill is something of a shared resource around the globe. The activities of one nation can undermine and potentially jeopardize all of the conservation efforts of another. Consequently, despite the many positives of environmental protection measures and the benefits of international treaties, the actions of one or even a few countries alone is not enough to ensure the continued existence of this rare and beautiful creature.

The hawksbill turtle was listed under the ESA as endangered in 1970. Critical habitat was designated in 1998 in coastal waters surrounding Mona and Monito Islands, Puerto Rico.

KEMP'S RIDLEY *(LEPIDOCHELYS KEMPII)*

The Kemp's ridley is the exception to almost every rule in the book. The smallest of all species, the rarest to spot and the most endangered, it was first submitted for identification by Richard M. Kemp in 1906, from whom it obviously derives its name. Growing to max lengths of only about two and a half feet and with an olive-gray carapace, the ridley weighs on average a comparatively light one hundred pounds. It has one major nesting beach on the Gulf coast of Mexico, nests predominantly during daylight hours, and is one of only two species of marine turtle to nest in what are called *arribadas*, the Spanish word for "arrival by sea" (see photo, opposite[3]).

The first Kemp's ridley ever recorded in Florida by the Turtle Patrol ambles over tire tracks on a busy beach in Ponce Inlet.

Think of an arribada as a sea turtle nesting convention. There is no one scientifically proven theory on what trigger these events—some working hypotheses include offshore winds, lunar cycles, and the release of female pheromones—but all at once, a mass of Kemp's ridleys will emerge from the tide to dig their nests in unison, dropping their eggs in a collective, synchronized mass. Waves of olive and gray rise from the shore, gather on a particular nesting beach from May to July and lay roughly one hundred eggs per female. This process is repeated two or three times throughout the season with eggs incubating for fifty to sixty days on average. On emergence, floods of one and one-half inch hatchlings weighing a meager ounce apiece will crawl from their nests to make their first attempt toward the rippling tides of the ocean.

Before entering pre-adulthood and more neritic waters to indulge on crabs, jellies, fish and an assortment of mollusks, Kemp's ridleys, like all other sea turtle species, are thought to swim into the open ocean offshore after emergence, where they can mature somewhere safe from coastal predators. This drifting period is thought to last about two years or until the carapace is about eight inches long. Once the turtle reaches this stage, it returns to coastal waters until full maturation, when it begins its mating and migration period.

For the Kemp's ridley, migration patterns are anything but well documented. While it seems that some males migrate annually between feeding and breeding grounds, others may not migrate at all, remaining in one set area of ocean and mating with females opportunistically encountered. Females, on the other hand, have been tracked going to and from Mexican beaches, then leaving feeding/breeding grounds for out-of-the-way foraging zones anywhere from the Yucatan Peninsula to southern Florida. Some females even remain in foraging grounds for months at a time, causing scientists to postulate that female Kemp's ridleys have goal-orientated migration habits, in place of the hypothesized wandering strategy thought of the Kemp's cousin, the olive ridley.

From Florida to New England, the Gulf of Mexico to the U.S.

Atlantic seaboard, adult Kemp's ridleys are distributed. There is also evidence of them occurring near Azores, waters of Morocco, and within the Mediterranean. As far as nesting is concerned, though, there is only one confirmed major arribada location, and it is in the state of Tamaulipas, Mexico, where almost ninety-five percent of worldwide Kemp's ridley nesting occurs. Arribadas do occur elsewhere but to a much lesser extent: places like Texas, even occasionally North and South Carolina, as well as the Gulf and Atlantic coasts of Florida.

Like most other sea turtle species, the Kemp's ridley population has endured a steep decline from where it was in its heyday. Take an unofficial home-video recorded in 1947, for instance. According to a written account on NOAA's (the National Oceanic and Atmospheric Administration) Kemp's ridley page on their website, this particular video, filmed on a nesting beach near Rancho Nuevo, Mexico, shows an estimated forty-two thousand Kemp's ridleys nesting in one single day. It also documents the mass harvesting of their eggs, with villagers seen excavating nests for hours at a time, and subsequent interviews suggesting that a possible eighty percent of these—thirty-three thousand nests, to put that in perspective—were stolen and taken away to local towns.

Just a mere twenty years after this video was filmed, the largest arribada on record was of only five thousand turtles. Worse, between the years of 1978 and 1991 only 200 Kemp's ridley were recorded nesting annually.

Fortunately, this particular sea turtle population does appear to be nearing the early stages of recovery. In the past ten years Kemp's nesting has seen a steady increase. The 2000 season, as an example, showed two thousand females nested at Rancho Nuevo. In the same location in 2001, a single arribada reported one thousand turtles. A massive jump occurred in 2003, when 3,600 turtles were estimated to have produced more than eight thousand nests. In 2006? The jump was even more incredible—12,143 nests in Mexico with 7,866 of those in Rancho Nuevo—marking a record number of nests recorded since monitoring of this species began in 1978.

Whether these variations illustrate a natural order of ebb and flow or the benefits of education and governmental involvement remains to be seen. In all objectivity, it's probably a combination of the two. Regardless, the taking of eggs from nesting beaches was, in fact, the Kemp's primary threat up until 1966 when the act was forbade under official federal law. Since then, this concern has become enormously less urgent, turning incidental capture in commercial fishing gear into their leading cause of mortality and otherwise adverse health effects. It just goes to show that when a species is left to its own devices, it is usually able to rebound outstandingly. Man's role in all of this is not to take a species by the hand and lead it into the future as much as it is allowing one room to showcase its own resiliency. Our job, assumedly, is not to play species piggyback but, simply, to stay out of the way. Which often can be harder.

The Kemp's ridley turtle was first listed under the Endangered Species Conservation Act on December 2, 1970, and subsequently under the ESA of 1973.

OLIVE RIDLEY *(LEPIDOCHELYS OLIVACEA)*

The only species of United States-resident sea turtle not to nest in Florida is the olive ridley. Like the Kemp's, this species forgoes solitary nesting in favor of arribadas, breaching the shoreline with hundreds, sometimes thousands of other turtles to partake in this ancient and communal ritual, one of the most perplexing and impressive of the natural world. Unlike the Kemp's ridley, olives are thought to be the most abundant sea turtle in the world, with an estimated eight hundred thousand nesting females annually. On Rushikulya Beach, India, for instance, an estimated two hundred thousand females have been documented nesting during one solitary arribada. The olive ridley may not be the largest or most migratory species in the sea, but their sheer density compared to that of their sister species is nothing

Olive Ridley[4]

short of staggering. It is not unheard of for a beach's nesting numbers to be so high that previously laid eggs are dug up by other females clearing the area to drop their own clutch.

Size-wise, the olive ridley is relatively unimpressive. It varies between twenty-two and thirty-one inches, weighing on average about 100 pounds. But in certain cultures—namely certain Asian and African societies—it is nonetheless considered sacred, the killing of one a highly forbidden act, which might help explain why its population has so far outshined that of peripheral species.

At around fifteen years old, slightly younger than other turtles, olive ridleys reach sexual maturity and begin nesting, usually once or twice a season, dropping around 100 eggs per clutch. Interestingly, although olives as a species do nest in arribadas, exceptions are still noted. Some olives, for unknown reasons, nest alone, and others, even more inexplicably, are mixed nesting strategists, joining in an arribada for one clutch then going it alone for the next. Suffice to say that this is just another trait to be added to the long list of sea turtle facts forcing scientists to simply shrug their shoulders. Hypotheses are offered, sure, but ultimately much of this behavior—the arribadas, the return to natal beaches for nesting, unpredictable migratory routes or inconsistent mating rituals—can only be ascribed to the yet-to-be-grasped mysteries of nature, the "it happens because it happens" rationale.

Though olive ridleys can be found offshore, inhabiting bays, estuaries and other coastal areas, they are mainly a free-roaming pelagic turtle. Breeding annually or sometimes bi-annually, they migrate from pelagic foraging grounds, to coastal nesting regions, back to pelagic zones to feed. They have been seen as far as 2,400 miles away from shore.

While foraging, olive ridleys will eat just about anything. They are omnivores, and so, feed on an array of grub from algae to lobster, tunicates to shrimp, diving to depths of 500 feet or more to snag any available benthic (life associated with the ocean floor) invertebrates they can find. And they'll do this just about anywhere, too—well, anywhere tropical. From the South Atlantic, to the South Pacific, to the Indian Ocean, olive ridleys will inhabit waters along the Atlantic coasts of Africa and South America. On the other side of the map, they'll occur from southern California to northern Chile, foraging opportunistically in these deep ocean tides.

Despite the many, many places they dwell, olive ridley arribadas occur only on a handful of beaches in the eastern Pacific and northern Indian oceans. In the Pacific, these nesting parties are thrown sometime from June through December on beaches in Mexico, Nicaragua, Costa Rica, and on one single beach in Panama. In the Indian Ocean, three beaches in India are host to arribabas. Solitary nesting, though, stands as the requisite monkey wrench in the nesting breakdown. "Lonely Hearts Nesters" (as your writers have come to call them) have been recorded across the globe in approximately forty different countries. If you were to fashion the olive ridley population into cliques, the solitary nesters are the outcasts, the wildcards. Where they'll nest next is anybody's guess. In high school terms, they were the ones who ate lunch outside, on some picnic table in the shade somewhere. They are the question marks.

Nesting habits aside, this is a yet another species of sea turtle that, although abundant in the scope of other turtles, is still very much at risk of extinction. According to the Marine Turtle Specialist Group (MTSG) of the Word Conservation Union (IUCN), since the 1960s, the olive ridley population has decreased an entire fifty percent, with degradation of nesting beaches, incidental capture in commercial fishing gear and ongoing harvest contributing to the loss. It's true, some nesting populations have shown growth in recent years, but the overall global loss is definitively greater than the gains. The most abundant of its kind, the olive ridley might also be the most exploited.

All along the west coast of Africa, nesting females and their eggs are regularly taken for consumption. Lack of data on specific beaches and trends make figuring these losses difficult, but the impact is surely devastating to the entirety of the eastern Atlantic population. In certain nesting populations in the western Atlantic, where statistical data is a little more rounded and reliable, there has been a measured eighty percent population reduction since 1967.

Long-term collection of eggs and the killing of nesting females remains the largest historical cause of death for this species, the arribada process setting up hundreds upon hundreds of easy, sitting-duck targets to those either in black-market trade or looking for a free dinner. This ongoing threat not only sandbags but can very realistically sabotage the recovery measures put in place to conserve this species. In Suriname, for example, despite explicit prohibitions against it, an estimated forty percent of olive ridley nests dug in 1995 were lost to harvesting. Imagine the toll that kind of continued abuse would take on a species after a decade, five decades.

But it's not just harvesting. Like all turtles, injury or death by commercial fishing gear is immensely impactful, as well. During the 1982-1983 nesting season, according to NOAA, over seven thousand carcasses washed ashore over a nine-mile stretch of beach on the Orissa coast of India. Between 1993 and 1999, that number skyrocketed to over 46,200, with a peak of 13,575 in 1997-'98. In fact, since 1999, over ten thousand carcasses a year have been documented on these beaches, most of which are attributed to shrimp trawl fisheries and gillnets. From incidental capture in this gear, an estimated sixty thousand or more sea turtles are killed along the coast of Central America every year, many of which are olive ridleys, creating a downward-facing trend that cannot possibly be withstood.

The olive ridley was listed under the ESA on July 28, 1978. The breeding populations in the Pacific coast of Mexico are listed as endangered; elsewhere the species is listed as threatened.

* * *

It's important to note that many of the regulations and behaviors listed for each species in the sections prior may also apply to others, as well—especially developmental habits and international guidelines and treaties. What changes are the nuances, in behavior, in exactly how fishing or protection rules affect each variety of turtle, and whether all countries respect the regulations in place. For more information on these subtleties, as well as deeper statistical information regarding nesting and population numbers, a plethora of helpful sources can be found in the Turtle Sources section in the back.

THE TOWNIES

In Volusia and Flagler Counties where the Turtle Patrol surveys, the most frequently found nests are laid by loggerheads, followed by green turtles. Leatherbacks also make sporadic appearances on these southeast beaches. Although Kemp's ridleys are extremely rare, from time to time they too are found.* Hawksbill nests appear to be non-existent, but because their tracks, nests and hatchlings so closely resemble those of Florida's most prominent species, the loggerhead, they could very well be nesting right under Patrollers' noses, and they just don't know it.

Differentiating between these nests and eggs can sometimes be difficult for the area's Turtle Patrol, but it's a talent that evolves in time. According to Danny Rutkowski, a ten-year member and sort of founding father of the Flagler sect, it usually takes about a year to become fluent in the language of turtle leavings, given the similar shape of eggs and track size, especially those of loggerheads and greens. Kemp's ridleys prove a bit easier in their clues. And for leatherbacks, determining species is never a problem, given the massive crop circle-like tracks they leave behind. "Leatherbacks are obvious," Rutkowski says, "Looks like someone drove a jeep out of the ocean, drove around in the dunes and drove back out." He can barely hold back a child-like smirk as he describes the enormous creatures.

* Kemp's ridleys are so rare in Florida that the first ever nest laid in the area wasn't recorded until 1996. The nest was documented by Beth Libert and nineteen-year Turtle Patrol member Eileen Day, who were lucky enough to witness the nesting in action on the beach of Ponce Inlet at 2:00 P.M. on May 15th. On emergence, state officials confirmed the species and the story made the front page of a local city paper. A second Kemp's ridley nest wasn't found in the same area until nearly ten years later, in June of 2005.

* * *

In the grand scheme of things, research and monitoring programs have been around a fairly short amount of time, and behaviorally, we still know very little about sea turtles as a species. This is important to remember when observing population trends and figures from any of the above sections. Given ten, twenty, fifty years' time, it all may be subject to great change.

Hatchlings on their way to the sea.

Danny & Ramona Rutkowski

Members of the Volusia/Flagler Turtle Patrol:
2000–Present

". . . get on the turtle telegram."

Danny Rutkowski was scheduled to come in to interview with his wife Ramona. But when he walked in alone, explaining that Ramona was working and couldn't make it, I almost feel embarrassed admitting the revelation that hit me like a ton of bricks: the Turtle Patrol *wasn't* her job. Danny, too, and all of the others—they patrol in their free-time, on the weekends or before an afternoon or graveyard shift. Most of them do this on top of working forty-hour weeks at schools and hospitals and construction sites and wherever else. They do this for nothing. They do this for fun.

On just looking at the guy you'd know he was a pro. He was laid back, spoke slowly, in fragments, and would infuse his answers to opinion-based questions with facts about nesting and track type, species differentiation and egg consistency. He wasn't doing it to brag or show off. He was doing it the same way a little league coach does when you ask him about his favorite ball player or what position he played when he was your age. He was doing it because he knew what he was talking about, and because it would make this book better.

[excerpt from interview]

Dan: We met Beth [Libert] thirty years ago. Moved up here, saw a thing in the paper looking for volunteers. It was Beth. Said, "Hey, let's go." Been doing it ever since.

How about your first sighting of an emergence?

Dan: Oh, it's exciting. I mean, the first time you see it, it's just incredible.

I'm sure there was a nice sense of satisfaction there too . . .

Dan: Oh yeah. And it's addicting. Just seeing those turtles come up . . . Catching one on the beach is probably the most exciting. And if you happen to be lucky enough to see an emergence, a first emergence, because what will happen is that first one will stick his head up. Like, I've walked up on a nest and I'll see that little head sticking out and I don't disturb him. I just watch him. When that first one makes a break—it's like a jailbreak! Then it's just . . . they pour out. You may have anywhere from fifty to a hundred just running for the water, just running to the water. Then while that's happening, again, you're keeping people away from them.

How many of those have you seen?

Dan: Dozens . . . Dozens . . .

Never gets old?

Dan: Never gets old.

CHAPTER 3

Threats

"For most of the wild things on earth, the future must depend upon the conscience of mankind."[5]

—DR ARCHIE CARR, FATHER OF MODERN MARINE TURTLE BIOLOGY AND CONSERVATION

NATURAL THREATS

With the exception of the occasional hungry shark, there remain very few natural predators to sea turtles once they reach adulthood. However, reaching adulthood can often be a long and arduous process rife with all kinds of peril. Scientists believe that only one in one thousand to ten thousand hatchlings actually survives to see its adult years.

On the beach, sea turtle eggs and hatchlings still *in situ* (in the nest in which they were birthed) are tantalizing meals for raccoons, crabs, off-leash dogs, and even ants. In the Volusia/Flagler region, though, those animals aren't the great threat they are elsewhere. Maybe it's because of the area's leash laws or the ample amount of food sources available for other predators, but in the Turtle Patrol's stomping grounds of northern Florida, the greatest threat to nested eggs stands the smallest of predators: the fire ant. Ghost crabs will also occasionally rummage through a nest, but the damage they do is minimal, consuming only one or two eggs per clutch, as

opposed to an *entire* clutch being obliterated by an army of fire ants. During storms, and especially during hurricane season, nests are also washed over by the tide or pulled out with it, depending on the degree of nest/beach erosion. And after an emergence, hatchlings are at risk from birds and crabs on the beach, and a number of predators in the sea, including a variety of fish, eel, and even dolphins.

But even after making it through the pounding waves and into the blue, turtles still aren't "home free"—not until they make it out of near-shore waters and into the open ocean. As small and feeble as they are at emergence, hatchlings cannot yet fight the tide and are thrown around like rag dolls in it, often drowned or carried back to shore only to repeat the first crawl process all over again. These "washbacks" are especially dangerous not only because they put hatchlings at a more drawn-out risk of predation, but because they fatigue these newborns to a point of exhaustion and complacency. After a few failed attempts toward the ocean, a hatchling might not have the strength or energy to try again and will end up stranded motionless on land, dry and more vulnerable than ever.

To the Rescue!

The Marine Science Center of Ponce Inlet, Florida, is a perfect example of the aid these organizations provide to hurt or dying animals on the coast. A professional rehabilitation facility, the MSC is able to provide medications, fluid therapy, blood work, radiology, and minor surgery to ill or injured sea turtles and sea birds.

As early as only March of 2009, over 170 turtles had been found sick, dead or emaciated on beaches from Jacksonville to Melbourne since the start of the year–which is more than three times the normal amount for that timeframe. Most of them were admitted to the MSC, kept in mini-pools, and some had to be tube-fed up to three times daily (a blended mixture of ground fish and other high-nutrient/high-calorie ingredients to promote weight gain), and many were rehabilitated and returned to the ocean as good as new.

For marine rehab clinics like these, no job is too big. In another Florida-based marine center, Sarasota's Mote Marine Laboratory, a five-foot-long, 787 pound leatherback was found stranded in Lee County and brought in for treatment (see picture below[6]). Because leatherbacks don't typically do well in confined spaces, the lab expedited its stay with them and released the mighty reptile back into the wild able to independently support itself after only a few days.

By definition, "strandings" isn't a term exclusively applied to hatchlings. Any dead or alive (albeit injured) turtle washed up on the beach is labeled a stranding. These might be turtles that experienced traumatic encounters with vessels, fishing gear or sharks, or they could be post-hatchlings or juveniles who simply couldn't fight the tide of coastal water. When strandings are found, the area's local Turtle Patrol is contacted and they transport the animal to the nearest marine center for treatment and rehabilitation.

Strandings and washbacks are, in simplest terms, basically the same thing. In its most commonly used form, a washback refers to a post-hatchling that has been out to sea for several weeks or months. This name was coined by Beth Libert back when these "tired little turtles" were cared for by her on her kitchen counters. The state classifies any hatchling between two to five inches to be a washback.

Tubs of washbacks from the 2007 hatching season.

Turtles are born with an innate inclination to swim until they reach the Sargasso Sea mats, which are far, far offshore. They'll swim for days, and on finally reaching the mats, they'll stop and hide there, opportunistically feed and grow to about half of full-grown adult size. At this juvenile stage of their lives, they become more free-roaming and lose the instinct to swim out to the deep sea if, by happenstance, they are pulled back in. During storm events in the fall, usually accompanied by large amounts of seaweed, these pre-adults are routinely washed to shore. Many people will mistakenly put these turtles back into the water when coming upon one on the beach, but this is more a problem than a favor. Without the instincts they had as hatchlings, the turtles do not know how to navigate the tide and won't make it out past the waves. These turtles must be given time to rest and recuperate before being taken out by a coast guard or marine center boat and released at sea, about ten miles out, far enough away from the tide so that they can swim back out to deeper water and more seaweed beds.

Adult and juvenile turtles are also often carried back to the beach during storms, cold weather or icy upwellings (an irregularly-occurring phenomenon where the normal warm water of the coast is blown out by strong and continuous southwest winds, then replaced by much colder water pushed up from the ocean floor). Depending on several circumstances, rehabilitated juveniles and adults are either brought out to sea for release, or let go from the shoreline.

In any of these classifications, dead variations can also occur: the post-hatchling, juvenile or adult turtle that is washed back to shore after a fatal run-in with a boat's propeller, disease, drowning or any other cause of mortality. Dead strandings are documented, recorded, and tagged with an "X" of orange spray paint on their backs by Turtle Patrollers. Then a sample may be taken for analysis before the carcass is buried underneath the sand. Freshly dead turtles and some rare species are salvaged for necropsies.

If you find an injured or sick sea turtle stranded at the beach or on a boat, please call the Florida Fish and Wildlife Conservation Commission at (888) 404-FWCC

HUMAN THREATS

"A McTurtle and fries, please": Consumption

Despite international laws and the spiritual importance turtles hold to a variety of cultures around the world, humans have consumed their meat for centuries and, unfortunately, in some cultures (Asian and Central American, especially) turtles and their eggs are still considered a common ingredient in drinks and things like turtle soup or turtle cake. In some regions, turtle eggs are considered aphrodisiacs. While the green turtle's meat is the most commonly consumed, loggerheads, ridleys, and sometimes hawksbills and leatherbacks are eaten, as well.

So prized a commodity are sea turtles in certain cultures, they are often consumed in conjunction with religious holidays. In Mexico, for instance, thousands of inland inhabitants journey to the Pacific shore the week before Easter to collect turtles and other seafood for the event. An estimated five thousand turtles are consumed during this time.

This kind of mass consumption doesn't occur exclusively outside of the States, though. In the U.S., eggs and turtles are snatched from beaches on a semi-regular basis. In July of 2008 an estimated 106 eggs were stolen in Flagler Beach, Florida by thieves said to be trolling the shore in the dead of night with a cooler in tow, digging up two egg-less areas before hitting pay-dirt with a third.[7] Around the same time in the same county, a couple of other incidents were recorded. This time, though, a man was caught while digging up a nest of eggs that were to hatch in only three to four weeks and was arrested. The fine for possessing turtle eggs is one hundred dollars for the first egg and sixty dollars for each additional. Poachers can even receive jail sentences.

Success story: Tortugeuro (translated as "Region of Turtles"), Costa Rica is the largest green turtle nesting site in the Western Hemisphere and, similarly, was one of the densest areas of turtle consumption in the world. During what started as a simple research mission in 1959, scientists and conservationists with the CCC (Caribbean Conservation Corporation) traveled there and began informing the villagers of the great threat sea turtles were facing. Continuing their research, they soon began working with the townspeople to find manageable life-style alternatives. Slowly, officials in the "Region of Turtles" realized that by changing their economy from one of consumption to conservation, they could effectively increase their economic potential through ecotourism. As a result, Tortuguero's green turtle nesting population saw a four hundred percent increase, and their economic plan is now used as a model for other small communities around the globe.

Live and Let Dive: Commercial Fishing

According to the CCC, more than 250,000 sea turtles are accidentally injured, captured, or killed each year by U.S. fishermen, with many of these incidents occurring during turtles' migration through fishing areas. With worldwide industrial fishing interests significantly larger than those in the U.S., it is probable that the global impact on this species at large is greatly more disastrous.

In fact, a study by researchers from Duke University's Nicholas School of the Environment and Earth Science, published in the research journal *Ecology Letters* in March 2004[8], revealed that more than 250,000 loggerheads alone and 60,000 leatherbacks are estimated to be entangled by commercial longline fishing on a yearly basis worldwide. Of these, it is estimated that tens of thousands are killed.

Sea turtles, especially loggerheads, are attracted to the bait suspended underwater by commercial longline vessels, often finding themselves caught on their thousands of individually baited hooks dangling from the surface, a kind of underwater minefield of glistening metal that can stretch as far as forty miles through the current.* Whether impaled by hooks or entangled in netting, these air-breathing reptiles suffocate while kept underwater by this gear, or they are injured or killed in its retrieval back onto a vessel's deck. This unintentional damage to sea turtles is referred to in the industry as "bycatch."

Bycatch most often occurs in four primary hotspots: the central and southern Pacific Ocean, the southern Atlantic and the Mediterranean Sea. Duke University researcher, Larry Crowder, even attests that more turtles are killed in the Pacific than actually nest.[9] Although their affinity for floating prey make loggerheads the most often affected by longline vessels, other species, such as leatherbacks, are also harmed, catching their giant flippers on hooks or tangling themselves in the line. Turtles are also sometimes struck by vessels' propellers or crushed when fishing apparatuses are machine-pulled back up onto a boat's deck.

---

* It is estimated in the Duke University study conducted in 2004 that longline harvesting fleets from forty individual countries suspended about 1.4 billion hooks in the recorded year of 2000—that's the equivalent of about 3.8 million hooks set every day.

Longline vessels set their bait at optimal depths and times to catch, usually, tuna or swordfish. Swordfish traps, however, snare ten times more turtles than those for tuna because of the shallower depths at which their hooks are set, and because they're cast during night hours, when turtles have a harder time identifying and avoiding these traps.

Although in the U.S. the National Marine Fisheries Service is working with the commercial fishing industry to implement technological changes to the types of gear used, including circle hooks and turtle extruder nets, researchers point out that only about two percent of worldwide longline fish landings are taken by U.S. vessels, thereby leaving a huge potential for death and injury elsewhere. According to the CCC, "the global fishing fleet is currently 2.5 times larger than what oceans can sustainably support."

Still, despite the United States' relatively miniscule percentage, at least a half dozen American conservation groups still believe they could be doing more. In April of 2009, six environmental groups including the Center for Biological Diversity, Defenders of Wildlife, the Caribbean Conservation Corporation, Florida Wildlife Federation, the Gulf Restoration Network and the Sea Turtle Restoration Project, sued the federal National Marine Fisheries Service (NMFS), claiming that they were in violation of the Endangered Species Act.[10] That NMFS has not yet determined the measures necessary to reduce incidental turtle capture is their crime, the six organizations claimed, citing high bycatch numbers from July 2006 to December 2008 as proof of their allegations.

The lawsuit was a power-play from these six groups to shut down longline fishing in the Gulf of Mexico until a better management plan was established (threatening more than thirty million dollars worth of industry from about one hundred mostly Florida-based boats). "The law clearly says the government cannot allow actions that jeopardize the existence of these species,"[11] Steve Roady, an attorney for Earthjustice, told a reporter.

But fishermen dispute the numbers brought forward in court, inciting a debate over the reality of bycatch figures as a whole.

"Not even close," a fisherman named Randy Baker told a reporter of the 2006-2008 bycatch figures presented in court.[12]

Another commercial fisherman, Woody Moore based out of Jacksonville, Florida, agrees. "I don't know what else they could expect us to do," he said, airing his concern over what a hiatus could do to the industry. "I've never in my life caught a dead one," he said. "And I've been fishing for thirty years."[13]

In the end, a temporary fishing ban was indeed enacted in the Gulf from mid-May to mid-October when sea turtles feed in its warmer waters.*

Bycatch is such a difficult issue to sort out not only because its figures are in debate, but because it can happen anywhere in the ocean. Because sea turtles migrate thousands of miles at a time, through vast areas of ocean and various territories, setting specific multi-national fishing agreements is a laborious process. One alternative to reformational international legislation that has begun its incorporation into the trade is the use of Turtle Excluder Devices, or TEDs, when fishing. A sort of compromise in the

*This was not the first time the fishing industry had been put on halt or suffered reductions in the name of turtle conservation. In 1999, higher than normal leatherback numbers were recorded in the Atlantic by the National Marine Fisheries Service and, under the Endangered Species Act of 1973, stricter laws were instated requiring the use of net and trawl alterations as well as a limit which "closed the ocean up" to about nine miles out, which is where most shrimp cluster. Although seen as good news for the turtles and its advocates, fishermen were angry. Headlines like the following from a May 14, 1999 edition of a local paper, the Daytona *News-Journal*, echo a strange transference of frustration, turning the animals themselves—not the city or the situation —into a direct object of resentment. The headline reads, "Turtles reduce shrimp harvest."

debate, TEDs are two-dimensional net inserts with large escape openings in shrimp and other trawl net fisheries. The method of fishing isn't changed with TEDs, only slightly modified.

Using larger circle hooks in lieu of traditional "J" hooks can also reduce the number and severity of sea turtle interactions with fishing gear. One study conducted by the NOAA Fisheries Pascagoula Laboratory and the NOAA Fisheries Miami Laboratory, in cooperation with Blue Water Fishermen's Association, showed that encounters with loggerhead and leatherback turtles were reduced by sixty-five to ninety percent by switching from traditional hooks to circle ones. Time and area closures have also been instituted in the scallop dredge fleet and various gill net fisheries, but this halt on the flow of business-as-usual commerce has the gross potential of starting an "us and them"-type mentality among conservationists and fishermen, activists and "everymen." It could very seriously affect the well-being of countless workers and peripheral industry businesses, as well, and should be used cautiously as an "if all else fails" alternative.

Turtle escaping trawl net through a turtle excluder device.

Trench Coat Commerce: Illegal Shell Trade

As was stated earlier, the intricate gold and brown patterns of the hawksbill's carapace have made it the sea turtle shell most coveted in international trade. Used to make jewelry and other luxury items, hawksbills are most frequently harvested and exported for these purposes from the Indian and Pacific oceans and the Caribbean Sea, where they are currently most especially threatened. Their population in these areas is estimated by scientists to have been reduced by some ninety percent in the last one hundred years. Other turtles may be slaughtered and processed as well, of course, to make leather products from their hide and beauty products from their oil.

The United States strictly forbids the buying, selling, or importing of sea turtle products, but tourists often unwittingly attempt to bring these items into the U.S. from the Caribbean, making sea turtle products—shell jewelry, turtle leather accessories and souvenirs—the most routinely seized contraband by international customs officials.

These trade regulations are enforced by the signing of 173 governments of the international agreement known as CITES (the Convention on International Trade in Endangered Species of Wild Fauna and Flora). CITES made the trade of turtle shells and oil illegal, but turtle mongering continues. The use of *bekko* (Japanese for turtle shell, usually that of the hawksbill or green) in Asian Pacific culture dates back more than 300 years. It was (and is) used to make combs that are an important part of the traditional Japanese wedding dress.

Friends Don't Let Friends Do Litter: Marine Debris

Mr. McGuire: "I want to say one word to you. Just one word."
Ben: "Yes, sir."
Mr. McGuire: "Are you listening?
Ben: "Yes, I am."
Mr. McGuire: ". . . Plastics."

—*THE GRADUATE,* 1967[14]

The Caribbean Conservation Corporation estimates that over one hundred million marine animals (all species) are killed each year by plastic debris in the ocean. Further, CCC reports that one hundred million tons of plastic exists in the world's oceans at this time, with some areas believed to be blanketed by five million square miles of plastic concentrated on the sea floor. To put that into perspective, five million square miles is roughly the area of the United States plus the entirety of India.

An estimated eighty percent of plastic in the ocean has been traced back to neighboring areas of land where it is washed away through storm drains and into rivers and streams, or flies from beaches and coastal areas where it is left by humans. Of course, plastics from careless boaters and beach-goers also add to this total. The Turtle Patrol collect what they can from their beach monitoring areas, but much escapes their grasp and ends up in the water and on the sea floor.

What are typical plastic products found in the ocean?

- Bags
- Bottles
- Balloons
- Product packaging
- Food containers
- Cigarette litter
- Fishing line
- Degrading buoys

All these can all be found easily. As any and all of these materials degrade, remnants of them are eaten by smaller species, allowing toxic additives entry into the water and, likewise, the food chain.

An anatomical feature of sea turtles' stomachs makes them especially susceptible to marine debris. Once taken into the turtle's mouth, plastic is prevented from being regurgitated by downward-facing spines in the throat, spines meant to keep down similarly gelatinous food sources like seaweed and invertebrates. Once trapped inside, these plastics prevent the turtle from breaking

"Just a tiny bit of what we find on the beach. The turtles look at this mess and think, 'YUM, what a tasty treat that is.' Awful."

Caption by Lori Ottlein

down food properly. This not only puts the kibosh on regular digestion but also leads to a condition rehab clinics refer to as "bubble butts." Materials blocking the digestive tract produce harmful gases during their decomposition, gases which eventually build up inside the turtle's body, causing it to float. This can lead to starvation and severe vulnerability to predators.

Dumpster Diving: Marine Pollution

While there are indications that the oceans surrounding the United States are less polluted than previously believed, pollution levels worldwide remain high. Even with federal sanctions in place and enforced since 1972 (the Federal Water Pollution Control Act), U.S. coastal waters are mired in great enough concentrations of hazardous material to threaten the well-being of marine animals across the board.

Oil drilling, as one of the seemingly more obvious examples, can uproot entire quartered off regions of the ocean floor and stain surrounding waters black if spillages occur. A startlingly high amount of the oil content found in our seas, however, doesn't actually come from rigs at all—but from inland runoff. Runoff through drains and rivers from nearby cities account for thirty-six percent of all marine oil pollution. Fertilizers also enter the water through storm runoff, causing an explosion of algal blooms referred to as *eutrophication*. This surge of excess nutrients slowly is absorbed by a rush of thriving plant life, depleting the reserves of surrounding oxygen in the water, suffocating marine life and creating oceanic "dead zones" in many parts of the world (the Gulf of Mexico, for example).

Whether by oil, fertilizer or decomposed plastics, recent studies indicate that the fatal turtle disease fibropapillomatosis may be linked to pollutants in the oceans. What many fail to understand about pollutants is that their toxins linger around long after the physical form has decomposed, continuing to contaminate aquatic life even after the original foreign entity is nothing but a faint, muffled echo in the sea. Because of the ocean's incredible size, it's easy to assume that these materials eventually dilute and disperse to negligible levels, but in reality they wreak havoc on the environment's natural equilibrium. Some toxins, for instance, grow more concentrated and more lethal as they make their way through the food chain, from small creatures, like plankton, that absorb these chemicals as they feed, to larger ones, like sea turtles, that eat the jellyfish that ate the plankton. The toxicity level increases incrementally through each form that it passes through—ocean, to plankton, to jellyfish, to turtle—creating an unnatural cocktail of chemicals that, once in the food chain, is forever in the food chain, becoming only more dangerous the more organisms it passes through.

In the case of coastal pollution, the long-time conservationist adage proves correct: the best way to combat these growing problems is to "Reduce, Reuse, and Recycle." Reduce oil consumption; reuse natural composts instead of chemical fertilizers; recycle plastic, glass and paper products. Our oceans today are still nowhere near pure, but the start of a visible upswing can definitely be observed. Due, at least in part, to the Clean Water Act and the Federal Water Pollution Control Act (FWPCA), maritime pollution has seen a modest downturn in the last thirty years.

The FWPCA demanded uniform, minimum standards on industrial water treatment and put limits on discharges from industrial facilities, deeming that all plants must adopt up-to-date pollution-control technologies. The Clean Water Act increased standards for waste treatment plants. About $125 billion was spent reforming public treatment works between 1972 and 1992, and the results have been massive. In Los Angeles, for instance, pollution input has been reduced by a whole ninety percent or more in the last twenty-five years, completely changing the face of the city's surrounding ecosystem.

If this trend continues elsewhere around the globe—who knows—in ten, twenty years, when other writers are writing other books on marine life and the threats it faces, maybe the requisite part about pollution might be a little closer to the History section.

"When the moon hits your eye . . .": Artificial Lighting

Because sea turtles lay their nests on beaches under the cover of darkness, and as far above the high water mark as possible, artificial lighting can disrupt the natural activities of both adult females and their hatchlings.

Throughout the world, and especially along the highly developed coasts of the United States, artificial lighting that shines onto beaches may discourage an adult female as she pulls herself out of the ocean to dig a nest. If the female becomes flustered or disoriented, she may return to the water to try another patch of beach or simply come back to the same one again later. If she is interrupted and turns back into the sea once more, resulting in yet another "false crawl," she may never deposit her eggs, instead unburdening herself of the load straight into the ocean and losing her entire clutch.

A green turtle's false crawl, returning to the water from whence it came.

When sea turtle hatchlings emerge from the nest, they appear to be lead in the direction of the brightest lights. Scientists believe this may be a natural instinct to allow the hatchlings to use moonlight on the ocean's surface to guide them into the water. Peripheral lighting that shines onto shorelines, though, can disorient the hatchlings, causing them to head away from the ocean and into alternative areas, waterless backyards or bustling highways. Hatchlings that have been disoriented by coastal lighting are run over by vehicles, eaten by animals, drown in swimming pools, or become too exhausted to ever find their way back home, through the sand and spray and into the tide.

The number of disorientations, though, entirely depends on the area surrounding these beaches. In Flagler County, Florida, for instance, according to Turtle Patrollers Dale and Peach Hench, the seasonal amount of disorientations on their route is negligible. "It's much more an issue in the south," they say, where condominiums and mega-strips line the coast. It also could be due to Flagler's higher dune systems, which block some of the light coming from the street. Regardless, the Henches record only one to two disorientations a season, they claim, a drop in the bucket compared to the numbers tolled in more commercial areas of Florida, like Daytona Beach.

Two Turtle Patrollers digging in a sewer drain to retrieve disoriented hatchlings.

The Gore Ultimatum: Climate Change

Though perhaps not exclusively a human threat, climate-related issues and the global warming controversy are concerns far too prevalent to simply be ignored. The melting of polar ice caps and the rising of sea levels, temperature change and the shifting of ocean currents, are phenomena that, if they occur as many scientists predict, could profoundly affect the sea turtle population.

If ice caps were to melt and sea levels rise, many beaches, wetlands and barrier islands where turtles nest would be submerged in the tide and lost. Since turtles rely greatly on their memories of emergence to decide where to drop their eggs, the loss of these landmasses could lead to fruitless searches for beaches that have all but disappeared, and countless clutches being lost to confused mothers who simply drop them in the water. Climate change could also affect water temperature which would, in turn, change the direction of ocean currents that are critical to migrating turtles—especially hatchlings, which transport almost solely on objects floating in the water for much of their early lives. Warmer waters would also mean quicker depletion of food sources. Coral reefs, an important food source among turtles, suffer from "bleaching" when water temperatures rise (see photo below[15]). This bleaching effect slowly kills off sections of the reef, resulting in fewer and fewer healthy and hospitable reef systems on our planet's face—which hurts, considering that almost half of the reef systems in the U.S. are already in mediocre condition.

Maybe most important to consider, though, is the possible change in air temperature. Air temperature obviously decides sand temperature, and since the temperature at which turtle eggs incubate decide their gender, any significant change could have serious repercussions on population variance. Scientists predict that increasing nest temperatures would result in more female than male hatchlings, creating a rift in the species' genetic diversity.

Interesting in all of the recent shifts and changes in temperature, marine environments and turtle populations are the discrepancies thereof. Since 2005, the Caribbean region has lost half of its coral, largely due to rising sea temperatures, and from 1997–1998 alone, mass bleaching was estimated to have destroyed sixteen percent of the world's reefs. As serious as these figures appear, certain regions have shown strong turtle numbers despite the dip in coral health. At the end of 2008, Volusia County, Florida hosted a record 649 sea turtle nests—the most since the advent of their Turtle Patrol monitoring program in the late eighties. Florida's Flagler County showed the same trend, when loggerheads dug more nests in 2008 than ever recorded. In Florida's Canaveral National Seashore, 3,637 loggerhead nests were spotted—the most since 2000. Even the extremely rare Kemp's ridley, which prior to 2008 had only two nests recorded in Florida's turtle monitoring history, dug a couple of nests in 2008 yielding sixty-two hatchlings.[16]

It is clear, 2008 was certainly a banner year in Florida's Volusia/Flagler region, hosting stable nesting and emergence numbers across the board. Still, even if these upward trends were to continue—which they did not—it would take decades to make up what has already been lost. Despite Florida's 2008 nesting records, turtle nests were still down statewide almost fifty percent to where they were only nine years prior, and despite further leatherback increases in 2009, Florida's loggerhead numbers in the same period were one of the lowest in twenty years. In fact, with the exception of 2008, Florida loggerhead numbers have dropped every year since 1998. Keeping in mind that ninety percent of the entire loggerhead population nest in only two places—Florida and Oman on the Indian Ocean—this is a worrisome prospect.

While Sunshine State nesting numbers seem to be in an odd limbo-like state of flux, other parts of the world are not only seeing steady climate-related declines, but are also beginning to take more drastic steps to ensure that balanced nesting seasons occur at all. In Playa Grande on Costa Rica's coast, the female to male ratio is through the roof, with nests being produced that are seventy to ninety percent female. At Junquillal Beach on the Pacific coast, it is too hot for eggs to hatch at all, and in these cases scientists have begun transporting nests to nurseries, seeing them through emergence and chaperoning them into the sea.

Whether the grimmer of these examples are signs of things to come or ecological anomalies still remains to be seen. What is important is that, despite political affiliation, we never take for granted our surroundings and take any and all reasonable measures to preserve them.

Dollar Store Chain Mail: Coastal Armoring

Coastal development has occurred everywhere, so much so that, in some areas, access to the world's beaches has been all but blocked by condominiums, hotels, homes and commercial buildings. In light of this urban build-up there are efforts to shield landowners' property against the natural effects of the wind and ocean, resulting in manmade structures like seawalls, bulkheads and rock revetments to protect against erosion. But the protection these structures provide is temporary and often quickens the rate of decay on neighboring sections of the beach. The same applies to jetties that are made to cut or expand inlets. Doing so causes the beach on one side of an inlet to erode much faster than the other, leaving an uneven coastline unable to repair itself.

Some stretches along the Turtle Patrol's native Flagler Beach, for instance, are covered in black tarp held down by massive rocks to keep the beach from taking over the streets above. The rocks in some areas, however, have washed away, leaving the surf and dunes a mass of crumpled black plastic. In other places where rock revetments are installed, the tide often extends all the way to the dunes, and every nest laid in these areas is at risk of washover and must be relocated by surveyors. The ratio of nests to false crawls where there are rock revetments is approximately three to one.

"Part of me just says . . . let the road go," says seven-year Turtle Patroller Barbara Ryan. "We have other roads. If this is nature's way of pushing the beach back further, why are we trying to stop it?" She cites streets that are eroding at rapid rates and others whose shoulders are already almost completely crumbled because the dunes underneath have washed out. "The road is *going* to wash away," she says. The only thing these walls, revetments and sheets of black are doing is delaying the process—but at the rest of the beach's costly expense.

Whichever the method, scientists have found that these armoring attempts often lead to less than optimum conditions for sea turtle nesting. Manmade structures alter the natural movement of beach sand, resulting in inhospitable nesting and breeding grounds. Some studies have even shown that fewer sea turtles emerge onto beaches with seawalls than onto nearby non-walled "natural" beaches. Further, those that do emerge in front of seawalls, will often return to the ocean without nesting, possibly because of turtles' impeccable sense of direction, their memories that are *imprinted* in their minds, possibly with magnetic maps, of the beaches on which they hatched, allowing them to return to that exact place (or, *almost* exact place) decades later to conduct their own nesting. If a seawall interrupts the natural flow of a beach's landscape or climate change-induced sea rise leads to the losing of their original birthing grounds, turtles can become disoriented and continue to search for locations that no longer exist.

Or, turtles that emerge onto walled beaches might have trouble finding an area near the dunes to nest that isn't supported by an underground boulder or a series of smaller rocks. The Turtle Patrol refers to mothers that run into these ruined nesting areas as "bumper turtles." You can see their tracks, these members say, come out of the water and then hit the rocks and crawl further along the dune, hitting the rocks again and then moving further along. They might even start to dig, but just a few inches below the

surface they'll hit rock and try a different location. The Turtle Patrol has come across handfuls of these tracks, sometimes extending hundreds of yards down the coast, the turtle trying and trying to find a place to lay, then giving up and going back out. If the urgency to nest gets great enough, mothers will jettison their clutch into the sea. If this were to happen every time one mother loggerhead, for instance, attempted to nest in a given season, it would equal about 400 to 500 endangered turtles being lost at Man's hand, a serious federal crime in any other circumstance.

Sea turtles utilize a great diversity of coastal communities, such as mangrove forests, lagoons, salt marshes, maritime hammocks, barrier islands, coastal strands and the beach and dune systems. The problem with beach fortification is that it affects all of these locales, reducing the overall health of the surrounding coastal ecosystem. If soil on a sea wall-free beach erodes in a storm, for instance, the exposed multi-layer root system of native plants—the sea oats, beach cord grasses, dune sunflowers and railroad vines—form a root wall in front of the exposed dune, acting almost as a sand trap, catching sand and eventually helping to rebuild and stabilize the dune. When walls are built, however, the entire landscape is skewed, not allowing the natural ebb and flow of beach reshaping to occur: erosion and accretion due to ocean currents, wave action and changes in the sea level. These seawalls impede longshore drift (the perpetual movement of sand along a coastline) and change the course of coastal winds, which constantly move sand along the shore to replace amounts that have moved or been washed away in other parts of the beach. Because sea walls don't move, don't build or wash away or contribute to new dunes forming elsewhere, the harmony of the beach's give-and-take with nature is disturbed, its interaction with the wind and rain, throwing off the forces which determine the slope, shape and size of any particular coastline.

SAND TRAP

If sea turtle populations enter into fatal declines, the nutrients that dunes would lose from their nested eggs would be enormous. Dune vegetation wouldn't grow as readily or as healthily, creating weak root systems unable to support dune structure. This would lead to quicker and more dramatic erosion, and, inevitably, the steady loss of beach volume. Every aspect of an ecosystem supports and relies on every other. If one cog in the system fails or falters, sooner or later the rest are bound to follow, tumbling over one another like a line of falling dominos.

The negative effects of beach armoring can especially be felt in Florida, where more than forty percent of the state's beaches are classified as critically eroded, caused by changes to the natural terrain.[17]

Another government-imposed beach change occurs when nourishment projects dump sand, which is often different than the existing sand, onto a shore in an effort to re-build a beach or dune system that has been eroded by various natural phenomena. While nourishment programs are usually preferable to armoring, they still aren't perfect. Because new sand is so often not the same make-up as the original that it's supplementing, unsuitable nesting places are known to spawn from these projects. Differences in sediment and compaction, for example, may cause a change in a turtle's digging behavior and nest site selection, or in the sand's incubation temperature and moisture content. These seemingly small alterations could decide whether or not a nest will survive until emergence. They can also play a hand in deciding the gender of the mother's clutch.

To add insult to injury, beach nourishment programs can also negatively affect near-shore reef systems. Dredging sand offshore then dropping it onto a beach can bury these habitats alive, especially in extra-wide man-made beaches, causing the loss of homes and food sources for hundreds of species, including sea turtles.

Our Own Worst Enemy: Human Beach Activities

Both daytime and nighttime use of beaches by humans threatens nesting sea turtles and their hatchlings. While coastal states ensure that residents and visitors have access to the beaches, more local

STRANGER IN A STRANGE LAND

Human activities away from the beach can decide the well-being of coastal ecosystems just as critically as can those in the sand. When forests were being cleared to make way for a new condominium complex in the Turtle Patrol's native Ponce Inlet, for instance, the habitat of many animals, including red foxes, was destroyed. Displaced from their natural forest environment, foxes soon began to roam the beach, digging up nests and stealing eggs, as well as eating newborn hatchlings. With the help of wildlife officers, Turtle Patrol president Beth Libert used female fox urine and powdered doughnuts to lure these displaced foxes into humane traps. Together they relocated the animals to woods miles away from the beach, where they might return to their normal dietary habits and lifestyle.

governments are realizing, sometimes only after being "convinced" by state and federal enforcement, that humans and sea turtles might need to be separated by some kind of buffer zone in order to protect the world's sea turtle populations.

Perhaps the most serious threat to sea turtles caused by human beach activity is disturbance to nesting females as they emerge from the ocean. Humans on the beach may unknowingly frighten a female from leaving the ocean in search of a nesting site. A person's very presence on the beach can sometimes be enough to put a turtle off or have her stop nesting completely to return to the ocean.

Driving vehicles on the beach can cause additional problems to these essential nesting grounds, too. Nests can be destroyed; beach sand can be compacted to a point of not being suitable for nesting; and hatchlings can encounter deep ruts or even be run over as they move toward the water. Night driving can create disorientation in emerging hatchlings, as well, whether in mistaking headlights for the moon's reflection off the sea, sand erosion, tire track impediments, or in simply blocking a turtle's path to the surf, extending its journey even further and increasing its probability of being caught by predators, forever remaining on the wrong end of the proverbial beverage.

Furniture and other items, such as kayaks, canoes, umbrellas, and boats, have also been known to confuse and, in some cases, entrap nesting females and their hatchlings.

* * *

No matter what the species, it is understood that sea turtles are beautiful and mysterious creatures, and some of the oldest on Earth. If given the opportunity to observe one, many will rightly wish to do so. What to keep in mind, though, are the regulations in place. Under the Marine Mammal Protection Act (MMPA), it is illegal and punishable up to a maximum $20,000 fine and a year in jail for engaging in the "take" of a turtle or its eggs. Just to watch is, of course, one hundred percent legal, but the conscientious observer will keep in mind a certain set of guidelines (see the inset "Turtle Viewing Guidelines").

Turtle Viewing Guidelines

Viewing Guidelines at Sea

Below is a list of NOAA's Southeast Region Marine Mammal & Sea Turtle Viewing Guidelines. Follow these basic codes of conduct to ensure that animals remain safe and undisturbed when encountering them in the ocean. These codes do not replace Federal requirements; they are for informative use only.

1. Remain a respectful distance from marine mammals and sea turtles. The minimum recommended distances are:
   - Dolphins, porpoises, seals: 50 yards
   - Sea turtles: 50 yards
   - Whales: 100 yards (500 yards for right whales)
2. Time spent observing marine mammals and sea turtles should be limited to one-half hour.
3. Marine mammals and sea turtles should not be encircled or trapped between watercraft, or watercraft and shore.
4. If approached by a marine mammal or sea turtle, put your watercraft's engine in neutral and allow the animal to pass. Any vessel movement should be from the rear of the animal. *Pursuit of marine mammals and sea turtles is prohibited by Federal law.*
5. Never feed or attempt to feed marine mammals or sea turtles. *Federal law prohibits feeding or attempting to feed marine mammals.*

More up close and personal encounters with sea turtles occur on the beach, and the handling of them and their eggs is conducted exclusively by trained and licensed members of the Turtle Patrol. If you are not a member of these ranks, the Caribbean Conservation Corporation has formed a list of mandates to keep in mind during encounters with turtles on land, while they are nesting or crawling to and from their nesting grounds. These types of encounters usually occur at night. The CCC list is as follows:

Viewing Guidelines on Land

- Do not walk on the beach with a flashlight or shine a light in the sea turtle's face. The light may cause the female to abort the nesting process, or other sea turtles nearby may be discouraged from nesting if there are lights on the beach.
- Do not take pictures using flashes. This high-intensity light can be even more disturbing than the flashlights.
- Stay clear and out of sight of the turtle until she begins laying eggs, otherwise you may scare her back into the sea.
- For your safety, stay away from the turtle's head. Sea turtles, especially loggerheads, have very strong jaws and can harm you if provoked.
- Do not handle the eggs or put any foreign objects into the nest. You can introduce bacteria or injure the eggs.*
- Do not handle or ride the sea turtle. In addition to being illegal, you may injure the turtle or cause her to leave without finishing nesting.
- Do not disturb tracks left by turtles. Researchers sometimes use the tracks to identify the type of turtles that nested and to find the marks of the nest.
- Do enjoy the experience, and remember it for the rest of your life.

* Eggs incubating in nests are extremely fragile. Even a slight shift in the angle at which they were set could separate the membrane from the embryo and make an egg sterile.

Barbara, Catherine, and Teresa Ryan

Members of the Volusia/Flagler Turtle Patrol:
2003–Present

Catherine: *"I* hate *mornings. The only time I ever get up is for turtles."*

Teresa: *"[And] it makes presents really easy—you just get something turtle-y. And every year you get more. Just turtles everywhere!"*

Catherine: *(laughing) "We're not obsessed . . ."*

Speaking with the Ryans was a lot of fun. They came in as a group, mother Barbara and her daughters, Catherine and Teresa, and laughed a lot at inside jokes, playfully ragged on one another, and reminisced about their favorite moments on patrol. Barbara is a long-time special educator in the Flagler County school system and Catherine is an elementary school teacher. They said they initially became interested in the Turtle Patrol after Teresa graduated from high school. They wanted to help her explore possible career options.

Trained by Danny Rutkowski, the Ryans didn't know anybody in the Patrol when they first started. Now, Teresa, the one in the family the others openly describe as "extremely shy," rearranges her work schedule so that she can survey multiple times a week. She also takes trainees out and, to her family's surprise, has no problem showing them the ropes.

"Because it's *for* the turtles," Teresa explains. "When you're at a nest evaluation and there's thirty people there and you have to stand up and explain what's going on—"

"That's love and devotion," her sister Catherine interjects. They all laugh. When the interview's over her mother looks at Teresa and says, "It's the most you've ever spoken!" And Teresa shrugs.

It's obvious that, for the Ryans, the Turtle Patrol is a family thing, a source of camaraderie shared not only by those three but also Barbara's youngest daughter and even her seventy-nine-year-old mother, both of whom come down for nest evaluations when crowds of 200 or so are expected, to help in education and "crowd control." After every morning survey, the Ryans go to the same local restaurant to eat breakfast together. Throughout the entire interview, whenever a question was asked, Barbara would look over and defer to her daughters. She would nod as they were speaking, smile and constantly pat Teresa, the daughter sitting nearest to her, on the knee.

[excerpt from interview]

Barbara: Our first season, the first nest Teresa found on patrol . . . when it was time for this nest to hatch, we would go out every night—nine, ten o'clock—and sit on the beach, by this nest, and wait for the baby turtles to come out.

Catherine: Not knowing anything of the signs of when a nest was supposed to hatch (laughs).

Barbara: We had a couple of clues, but—

Catherine: We didn't really know what we were doing at that point.

Barbara: And the one night it was raining and, you know, we sat there because we were happy sitting there.

Catherine: 'Till, like, two or three in the morning with umbrellas.

Barbara: And as we're sitting there, I look down the beach and there's a mamma turtle coming up to nest. So while we're waiting for one nest to hatch, we got to see another one nest. You know, you just can't beat that for entertainment.

Especially in your first year to see that. And you're with your family, and I'm sure that has a lot to do with it, also.

Barbara: Oh yeah

Teresa: We were just like, 'Holy crap!'

Barbara: It's the Creature from the Black lagoon! It's a *sea* creature. It's supposed to be *in* the water. And it's coming *out* of the water (Catherine is cracking up).

Teresa: We didn't know what it was. It was just this big black shape coming out—

Barbara: *Blob!*

Teresa: And we were like, '. . . That's a turtle.' And then we're trying to get up and we're trying to figure out—

Catherine: Remember hiding behind the walkover?

Teresa: We were hiding underneath the walkover because until they actually start dropping eggs, you can scare the turtle away. And so, we were trying to get close enough to see if she had started dropping, to see if we could get closer. But we didn't want to scare her away so we're—

Catherine: We're calling Ramona and Dan on the phone!

Barbara: On our bellies . . . under the walkover (all laughing now).

Catherine: And I think mom (stops laughing to catch her breath) stayed back at the walkover and we did the *army crawl* over to [the turtle], going over behind her.

So would you say that was your best memory?

Teresa: (all nod heads) Because it's not very often you get to see the nesting.

Barbara: But we've seen a couple since then. But it's not the same as your first.

Teresa: Our first season, we saw—like, there were five turtles nesting, which is *really, really* rare.

Do you remember what species the first was?

(In unison): Loggerhead

Catherine: And it was the sheer power she had. Because once she started to cover up her eggs, we got pretty close. . . . Looking back, we probably shouldn't have [gotten that close]."

Well, you didn't know any better.

(In unison): . . . Yeah we did! (laughs)

Images from Teresa Ryan's Scrapbook

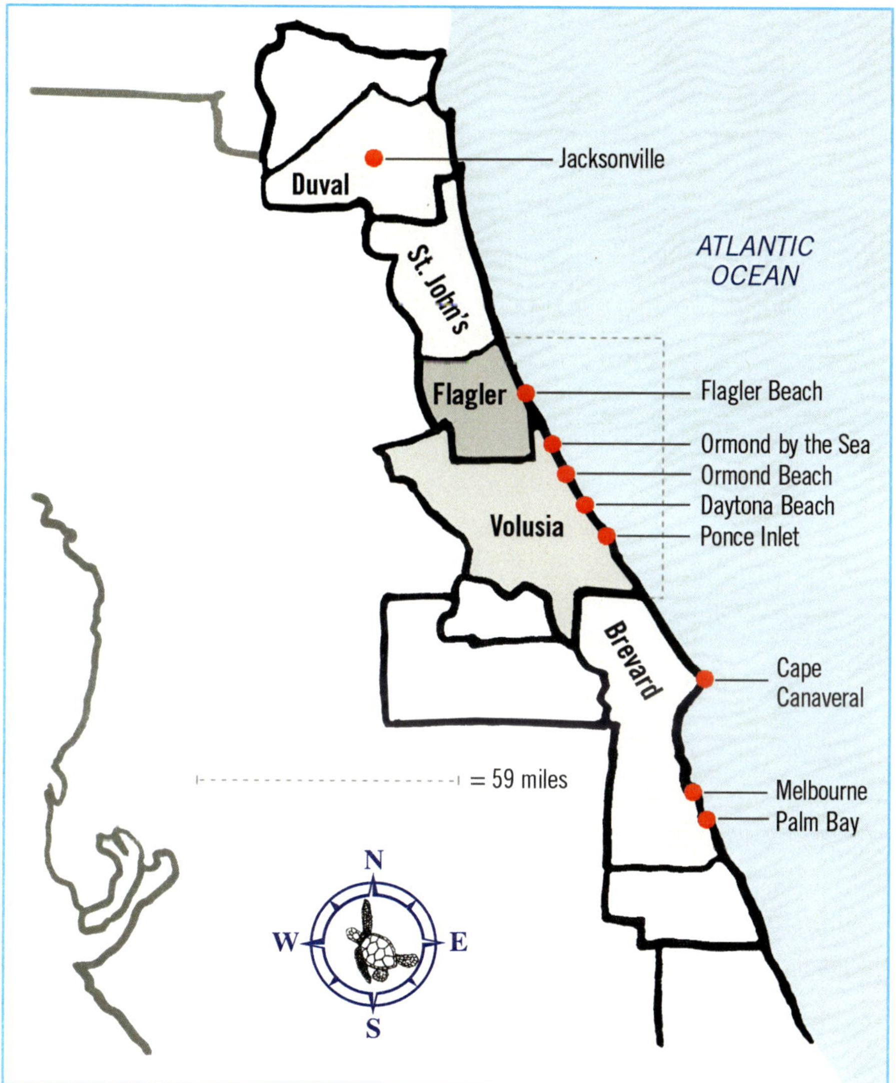

Map of Eastern Florida: Flagler/Volusia County Region

CHAPTER 4

The Beginning, Revisited

"A guy by the name of Blair Witherington said it best: 'They are like artists. Different artists have different strokes—you can recognize their paintings. Different turtles have different strokes.' You can tell which kind, what kind of turtle has come on the beach just by the way they crawl. Because you don't see the turtle, they're gone. They've come and gone. All you see is the tracks."

—DANNY RUTKOWSKI, TEN-YEAR MEMBER OF THE TURTLE PATROL

BIRTH OF A PATROL: METHODOLOGY AND THE EARLY YEARS

In the beginning there were two.

Richard and Joyce Durrbeck knew little about sea turtles when their daughters came to them one day for help on a school science project. This was before the Turtle Patrol was the Turtle Patrol. It was before the daily beach surveys and the state permits and the tee-shirts emblazoned with the group's insignia. Back then, in a late summer in the early 1980s, it wasn't yet about turtles or environmental conservation. It was about the Durrbecks' children, Sharon and Stacy. It was about them needing help.

To learn more about the subject of the girls' project, the Durrbecks took a ride to The Turtle Lady in Ponte Vedra Beach. The Turtle Lady was a non-profit organization inspired by a woman named Ila Fox Loetscher to conserve the critically endangered Kemp's ridley from extinction. Ila was an ex-pilot from Texas in the years just before the Great Depression, the first licensed female pilot in Iowa's history, and before turning her attention to sea turtles, started with her friend Amelia Earhart the country's very first women's flying organization, the 99 Flying Club, in 1931.

It wasn't until years later in 1966 that Ila became a highly contributing member of the conservationist community, joining an activist movement called Project Ridley that made yearly pilgrimages to Rancho Nuevo, Mexico, to relocate as many turtle eggs as possible back to the beaches of South Padre in hopes of re-establishing the area as a thriving nesting beach. She was always out in the field, conducting relocations, educating her neighbors, and more and more her efforts began to gain attention from tourists, residents and environmental activists alike. Soon she became the go-to person in her community for any and all things turtle, and even received a commendation by Jacques-Yves Cousteau and his Cousteau Society in a 1981 edition of the *Cousteau Almanac*, bestowing on her the title of "Wavemaker." Once again, on Ila's eightieth birthday, she was recognized, receiving a personal letter from the Cousteau Society lauding her lifetime achievements. In part, the letter read:

> "Your work with sea turtles has helped bring the plight of this enigmatic creature more into focus for scientists and laymen alike. I recently spoke to Dr. Archie Carr who echoes strongly these feelings."[18]

From those beginnings, Ila Loetscher became "the Turtle Lady," a title that would later be handed down and inherited by other women in other states and towns, like Beth Libert, who dedicate their lives to working with turtles. It was trailblazers like Loetscher who brought the sea turtle's plight into the public eye—before her kind, few knew or scarcely cared about the troubles facing the slow-moving shelled creatures crawling about their beaches. And the Durrbecks, visiting Ponte Vedra's Turtle Lady to better understand how to engage their children, were moved by the weight of that history. At the end of the day, when the family stood witness to the release of loggerhead hatchlings into the ocean, they were officially inspired. Later, they would visit the House of Refuge museum of the sea on Hutchinson Island in Stuart, Florida, to observe the injured turtles that were kept in their holding facility, as well as the baby hatchlings in their "Head-Start Hatchling Program."

With this newfound excitement, the Durrbecks commissioned Ross Witham, who was the sea turtle coordinator for the Department of Natural Resources (DNR) at the time, to help them apply for a permit to survey and relocate nests. The permit would be issued through the Florida Fish & Wildlife Conservation Commission, which derives its authority from the U.S. Fish & Wildlife Service. Long story short, they acquired the permit, and because nobody else in the general vicinity conducted turtle patrols, the Durrbecks became responsible for the entire area from Ponce Inlet north to the Flagler County line.

Back then, a mere thirty years before this book was written, it was truly a different time in the sphere of turtle research and protection. There were no local ordinances protecting sea turtles; vehicles were allowed to roam the beaches; coastal lighting was not regulated; and nests were openly vulnerable to all sorts of damage, incidental or otherwise. To stand any kind of substantial run at survival, every nest recorded had to be scooped out and carried to a safe, off-site location to incubate. The Durrbecks' garage became this location.

For a long time this was common practice: traveling out to a nest with a Styrofoam cooler in tow, lining the inside with enough sand to keep the eggs still and at a consistent temperature, filling it up with a clutch dug from the ground and transporting it back to the garage for safekeeping. Once accustomed to the work, Richard and Joyce were able to collect more and more eggs along the coast, eventually filling their home garage with lines upon lines of stacked coolers, all warm and ever so slightly vibrating with the muffled pulse of hundreds of quietly pulsating hearts.

Meeting Beth Libert that day on the beach, to the Durrbecks, was a godsend. Although at first Richard was admittedly suspicious of her commitment ("Everybody says they want to help," he told her, "but not many are willing to do the work."), Beth soon proved herself, quickly becoming as essential to the rapidly-forming Turtle Patrol foundation as they were. The Durrbecks knew how difficult it would be to cover their allotment of landmass alone, and after a couple of years of mostly solo work, Beth was added to their state permit and given the fifteen-mile area from Volusia's Main Street Pier south to the Ponce Inlet jetty to monitor. In her first year, she excavated eighteen nests, stored eggs in her garage, recruited a few volunteers and, when eggs were ready to hatch, would load the neighborhood kids into her pickup truck and take them to the beach for the big release. This was an event they all looked forward to.

Soon there was a system in place, a "how this is done." In 1984, more players were added to the ranks. Marye Mar-

A young Beth Libert brings a cooler of newly hatched baby turtles to the beach for a release.

shall and her husband Frank signed on and took over the ten miles of New Smyrna Beach, south from the Ponce de Leon Inlet to Canaveral National Seashore. This addition guaranteed the monitoring of the entirety of Volusia County by volunteers. After reaching this landmark, the group decided it was time to make a name for themselves. They chose the Volusia Sea Turtle Society.

And so the society continued to grow, from two members to three, three to five, five to over a hundred, most of them individuals who came across Beth or the Durrbecks on the beach and stopped to see what they were doing. Most new members signed on not from direct solicitation but from word of mouth and minor newspaper publicity. They were usually interested people looking to be a part of something bigger, eager to partake in an activity exclusive to their home state, whether they'd lived there their whole lives or just recently moved. It was a Florida thing. You couldn't dig eggs from the ground and watch them turn into turtles in somewhere like Wisconsin or Nevada. Where else could you wake up, set out for the untouched morning beach, its waves rolling in on the shore like a slow, tired breathing, and watch the sun rise, feel it warm the air around you as you scour the sand for life that needs protecting? Where else but Florida could you experience the breezy gold of a beach at dawn, feel the rubbery white of a tiny egg in your palm and know that, because of you, it might live to see one-hundred years? Where else but Florida could you be this much a part of the shoreline?

Later, the Patrol would pick up so many new members that the real challenge would become training all of the newbies while holding onto the quality volunteers. Many of the veterans were (and are today) school teachers and college students, and many would quit or go on hiatus once the school season started and they'd have to return to class or work; others would tire of the hard work and long, thankless hours. Several years later Beth Libert would become the Patrol's president, or "turtle mother," as some affectionately refer to her. Years after that, in 2008, the Patrol would take over part of St. Johns County, as well, an extension that became especially taxing when Beth realized she'd have to re-structure the entire division and re-educate all of the existing members on routine protocol. In that region, the whole lot of volunteers weren't adequately trained to properly identify, excavate or relocate nests. The fifty volunteers Beth was told she would have to monitor St Johns County ended up pairing down to about ten capable people. After that, mandatory meetings in April for new volunteers and regular monthly meetings were instituted in each county in order to keep everyone organized and on steady track. These meetings are held in community centers or Turtle Patrol offices in coastal policy centers.

It wasn't until Richard Durrbeck passed away in 1992 that Beth Libert stepped up and took over his permit for the twenty-five miles north of Ponce Inlet. She wrote about this early period, the formative years of her and the Durrbeck's organization—the "rise" of the Turtle Patrol, if you will—in the very first edition of the Patrol's newsletter, the *Turtle Tracker,* published in 1995. Her words are commemorative and nostalgic, a sort of tip of her hat to what used to be, and *who* used to be. Her final words directly reflect the latter sentiment:

"We miss you Richard!"

WHO PATROLS THE TURTLE PATROL?

For roughly five years Beth and the Durrbecks employed the method of in-cooler artificial incubation for turtle eggs. This was the unquestioned norm, until the Florida Department of Environmental Protection stepped in, proving it an unhealthy method and, in some cases, potentially incurring more harm than good, considering how extremely fragile turtle eggs can be, and dependent on nest temperture.

Things began to change for the Turtle Patrol. In 1989, instead of taking nested eggs back to volunteers' homes, they now by law had to be transported from Ormond Beach north to Ormond by the Sea, and from Daytona Beach south to Ponce Inlet, where they were to be buried in designated areas of sand, or "hatcheries," for natural

emergence. More than three thousand eggs were buried this way in clearly marked areas along the dunes of Old Timers Beach, the area's most natural and secluded setting.

Over time, the Patrol was becoming more legitimate and official. No longer just a couple of well-meaning enthusiasts trolling the beach for nests, the Turtle Patrol had to adapt to its growth and begin adhering to a more extensive list of federal regulations. The process of nest relocation lasted only for a couple of years before the government intervened once more. Then, eggs were no longer allowed to be moved from their nests at all unless they were in a highly compromised position (i.e.: buried below the high-tide line and in immediate danger of being washed over*; or in some cases, if a major storm, like a hurricane**, is imminent, exposed eggs could be salvaged and relocated to higher dry ground). New common practice detailed that all nests, barring those in dire circumstances, must remain *in situ* (where they were laid by the turtles), and sectioned off with a square of four wooden stakes lined with ribbon barrier and a sign, alerting beachgoers to the line they are legally not allowed to cross.

Marked nest washed over by pre-storm high-tides.

In conjunction with these new standards, every volunteer on patrol also began keeping a daily log of their activities. Every nest had to be recorded and, in Volusia County, these reports and logs are still today sent to the county daily for documentation. On top of fencing off each one, marking the stakes with the appropriate information (an assigned nest ID number, the date laid and volunteer initials) and using the appropriate ribbon (ribbon color is

* No one knows for sure just yet exactly why turtles, after so many centuries and such an established instinct to nest on land, will sometimes dig their nest chambers below the high-tide line where their eggs will almost undoubtedly be washed over or pulled out. Danny Rutkowski simply attributes the phenomenon to the mother turtle being a "lazy girl." Others liken the scenario to a taxi delivery. You can speculate, they say, that the turtle's laying process was simply too far along, and it was just *time* to lay the eggs—almost like she couldn't hold them any longer. This was her stop.

** Hurricane season remains one of the rare threats the Turtle Patrol can do little to guard against. Extending months through the summer, this prolonged storm season can claim hundreds upon hundreds of eggs nested on the beach. Surveyors might relocate eggs to higher ground or take them offsite, but many will inevitably be pulled out to sea, washed over, and drowned. "You spend all season saving these nests and marking them all off," Teresa Ryan explains, "[then] they're all gone." In preparation for some storms—like Hurricane Floyd in 1999—mandatory nest evacuations are ordered.

assigned by month to assist nest watchers: May—pink; June —blue; July—orange; August—green; September—pink again), each volunteer also began recording the weather, surf temperature, any violations observed, the location of all nests found, and any actions taken while surveying. Jobs became more rounded. Through measures to protect the turtles, what the group essentially wound up becoming were patrollers of the entire beach. Their daily log has been slightly tweaked throughout the years but the skeleton of the one formed in 1997 remains the same. The log used today looks like this:

VOLUSIA TURTLE PATROL DAILY LOG 2008

Date__________ Start Time_________ End Time_________ Volunteers __________________

Weather ____________________ Surf Temp. _______

Crawl, Event or Violation	Address & City Initials	Action Taken or Comments
FLAGS (circle when put up) Dunlawton- FL Shores- Perry's Beach Street Main St. - University Cardinal - Granada	Example: Famous Motel 2505 S. Atlantic Ave. DBS	Example: Car on beach at 6:15 am. Called Sheriff's office to dispatch Beach Patrol.

A separate form must be filled out and submitted whenever a nest is relocated. This one is a little more in-depth. It defines not only the time, date, GPS (Global Positioning Satellite) location and reason for a relocation, but also contains a follow-up portion that is filled out upon the nest's emergence. This form looks like the one below and on the facing page.

Nest ID # ________________

RELOCATION INFORMATION

Relocation Date: ____/____/____ **Relocation Time:** ______________

# of Relocated Eggs: ________ **# of Eggs Broken:** ________ **Clutch Depth:** ________cm

Relocation Location: ____

Number Street City ZONE Facility Name/Landmark

Shore Type: ☐ Dune ☐Sea Wall ☐Revetment

Distance of relocated eggs to dune stake ___________.________ m

GPS Coordinates N29______________________ **W081**______________________

Remarks: ____

NEST FATE INFORMATION **DATE OF EVENT** ____/____/ 08

☐ Excavation Attempted but Clutch Not Located ☐ Vandalized/Poached ☐ Washed Out

☐ Nest Not Excavated Reason: ______________________________________

☐ Depredated By (Check One) ☐Raccoons ☐Foxes ☐Dogs ☐Armadillos

REPRODUCTIVE SUCCESS DATA

1st Hatchling Emergence Date: ____/____/ 09 **Observed by** ______________________

Other Emergence Dates______________________________

Nest Excavation Date:____/____/ 09 **Nest Excavated By:** ______________________

Hatchlings Disorientation(s): ☐Yes ☐No **If Yes, Dates:** ______________________

Excavated Before 72 hr: ☐Yes ☐No **Reason:** ______________________

Egg Chamber Contents: Include only those items found inside the nest (not on ground above nest)

Live Hatchlings	Dead Hatchlings	Pipped Live	Pipped Dead	Unhatched Eggs		Eggshells (in situ nests only)
				Whole	Damaged*	

* Eggs damaged by predators, roots, nesting females, or during relocation.

Present in Egg Chamber? Roots: ☐Yes ☐No **Fire Ants:** ☐Yes ☐No **Ghost Crabs:** ☐Yes ☐No

OVERWASH DATES: ____

REMARKS: ____

____

Relocation Form, side 1

FLAGLER COUNTY BEACHES

Survey Date: ___/___/___ **NEST ID #**__________
(Do not number false crawls)

Observers: ______________________________

Survey Area (Circle One): FB, FN, FS, ML

Type of Crawl: ☐Nest ☐False Crawl **Interpretation:** ☐Positive ☐Reasonably Certain ☐Questionable

Species: ☐ Loggerhead ☐Green ☐ Leatherback ☐ Kemp's ridley

Abandoned Digs: ☐No ☐Yes # of ABPs only ______ # of ABPs w/ AEC ______

Original Crawl Location: ______________________________
Number Street City ZONE Facility Name/Landmark

Shore Type: ☐ Dune ☐Sea Wall ☐Revetment **Distance: of Clutch (Before Relocation) to:**

☐Landward (West) ☐Seaward (East) of Toe of Dune **Dune Stake:** ____ . ____ m
***(in situ* nests only)**

☐ Above ☐ Below Most Recent High Tide Line **Toe of Dune/Seawall :** ____ . ____ m

Obstacles Encountered During Crawl: ______________________________

Clutch Located ☐Yes ☐ No

CRAWL SKETCH
(MUST show dune, CZ, and recent high tide line)

Nest Relocated: ☐Yes ☐ No
(If Yes, complete reverse side)

Why: ☐ Tidal inundation ☐ Runoff

Comments (Date of Crawl Only):

GPS Coordinates
N29
W081

Relocation Form, side 2

But there was more paperwork, still. Disoriented turtles had to be reported, as did stranded ones. Keeping such diligent records on any and all turtle encounters was and is the only way for cities and federal officials to track population numbers, measure how serious a threat coastal lighting is in their communities and learn how effective are their endangered species regulations. By law, the Turtle Patrol is required to provide sea turtle stranding and disorientation reports to the National Marine Fisheries Service (NMFS). From this information, NMFS provides a weekly update from all states in the southeast region and sends them to various

Permit Holder Initials — Year — Month — Day — Dis. # by Day — County Code

FWC MARINE TURTLE HATCHLING DISORIENTATION INCIDENT REPORT FORM

If you have any questions please contact FWC at the Tequesta Field Laboratory (561) 575-5407 or in Tallahassee (850) 922-4330

Turtle Permit #: ______________ Date of Incident: ______________

Observer's Name: ______________

Telephone (include area code): ______________

Location of incident (address of source, beach name and/or nearest landmark): ______________

City and County: ______________

Local nest ID# &/or zone nest was located in: ______________

Address/landmark hatchlings disoriented towards: ______________

Was a probable/possible lighting source identified? YES ____ *NO ____

If so, what type(s) of light(s) were identified? (please circle)

parking lot — street light — condominium (interior)

dune crossover — single family home (interior) — condominium (exterior)

restaurant/bar — single family home (exterior) — sky glow/urban glow

pier — other: ______________

*If not, why?: (please circle) Too many lights present to determine — No possible lights observed

Describe lighting source(s); include number & type of lights observed: ______________

Incident was documented during (circle one): morning survey — night survey

Was this a caged nest? YES ______ NO ______ If yes, what type of cage? ______________

Was a temporary light barrier used (i.e. Silt screen)? YES ____ NO ____

Was this a relocated nest? YES ____ NO ____

Was the incident photographed? YES ____ NO ____

Was the nest located? YES ____ NO ____

Was the nest excavated? YES ____ NO ____

If yes, how many hours after emergence? ______________

	LOGGERHEAD	GREEN	LEATHERBACK	UNIDENTIFIED
No. OF HATCHLINGS DISORIENTED				
No. OF HATCHLINGS FOUND DEAD				
No. OF HATCHLINGS FOUND ALIVE				
No. OF DISORIENTED HATCHLINGS REACHING WATER				

Waterline

Sketch

Additional comments (please elaborate and use back if necessary): ______________

Was local authority provided a copy of this report? YES ____ NO ____

If yes, please indicate person and city/county/state department report was copied to: ______________

Signature of Observer ______________ Date ______________

DNR FMRI 33-714 Revised 6/29/92, 11/12/96, 9/97, 1/99, 3/01, 11/01 law enforcement personnel and protected species personnel. All reported strandings should be photographed and the FWC should be notified within twenty-four hours of reporting.

Today's forms for strandings and disorientations look like the one below and on the facing page.

SEA TURTLE STRANDING AND SALVAGE NETWORK – STRANDING REPORT

OBSERVER'S NAME / ADDRESS / PHONE:
First ______ M.I.____ Last ______
Affiliation ______
Address ______
______
Area code/Phone number ______

STRANDING DATE:
Year 20☐☐ Month ☐☐ Day ☐☐
Turtle number by day ☐☐

Coordinator must be notified within 24 hrs; this was done by ☐ *phone* ☐ *email* ☐ *fax*

SPECIES: (check one)
☐ CC = Loggerhead
☐ CM = Green
☐ DC = Leatherback
☐ EI = Hawksbill
☐ LK = Kemp's Ridley
☐ LO = Olive Ridley
☐ UN = Unidentified
Check Unidentified if not positive. Do Not Guess.

STRANDING LOCATION: ☐ Offshore (Atlantic or Gulf beach) ☐ Inshore (bay, river, sound, inlet, etc)
State ______ County/Parrish ______
Descriptive location (be specific) ______
______
______
Latitude ______ Longitude ______

Carcass necropsied? ☐ Yes ☐ No
Photos taken? ☐ Yes ☐ No
Species verified by coordinator?
☐ Yes ☐ No

CONDITION: (check one)
☐ 0 = Alive
☐ 1 = Fresh dead
☐ 2 = Moderately decomposed
☐ 3 = Severely decomposed
☐ 4 = Dried carcass
☐ 5 = Skeleton, bones only

FINAL DISPOSITION: (check)
☐ 1 = Left on beach where found; painted? ☐ Yes* ☐ No(5)
☐ 2 = Buried: ☐ on beach / ☐ off beach;
carcass painted before buried? ☐ Yes* ☐ No
☐ 3 = Salvaged: ☐ all / ☐ part(s), what/why? ______
______
______
☐ 4 = Pulled up on beach/dune; painted? ☐ Yes* ☐ No
☐ 6 = Alive, released
☐ 7 = Alive, taken to rehab. facility, where? ______
______
☐ 8 = Left floating, not recovered, painted? ☐ Yes* ☐ No
☐ 9 = Disposition unknown, explain ______
______
If painted, what color? ______

SEX:
☐ Undetermined
☐ Female ☐ Male
Does tail extend beyond carapace?
☐ Yes; how far?______ cm / in
☐ No
How was sex determined?
☐ Necropsy
☐ Tail length (adult only)

TAGS: ***Contact coordinator before disposing of any tagged animal!!***
Checked for flipper tags? ☐ Yes ☐ No
Check all 4 flippers. If found, record tag number(s) / tag location / return address
______
______
______

PIT tag scan? ☐ Yes ☐ No
If found, record number / tag location
______
______

Coded wire tag scan? ☐ Yes ☐ No
If positive response, record location (flipper)
______

Checked for living tag? ☐ Yes ☐ No
If found, record location (scute number & side)
______

CARAPACE MEASUREMENTS: (*see drawing*)

Using calipers		**Circle unit**
Straight length (NOTCH-TIP)	______	cm / in
Minimum length (NOTCH-NOTCH)	______	cm / in
Straight width (Widest Point)	______	cm / in
Using non-metal measuring tape		**Circle unit**
Curved length (NOTCH–TIP)	______	cm / in
Minimum length (NOTCH-NOTCH)	______	cm / in
Curved width (Widest Point)	______	cm / in
		Circle unit
Weight ☐ actual / ☐ est.	______	kg / lb

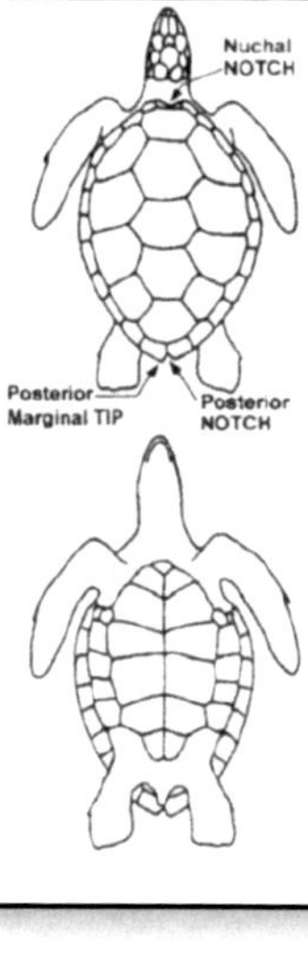

Mark wounds / abnormalities on diagrams at left and describe below (note tar or oil, gear or debris entanglement, propeller damage, epibiota, papillomas, emaciation, etc.). **Please note if no wounds / abnormalities are found.**
______
______
______
______
______
______

By this time in the late-1990s, what had become the Volusia Turtle Patrol, Inc. had assembled a great deal of members, established a leader and taken over the permit for Flagler County, further legitimizing the organization and expanding its name accordingly. They began fighting against beach driving to protect turtles and their nests —lawsuits were even brought to court by a woman named Shirley Reynolds on this very subject—and later rallied against nighttime coastal lighting. Today, every nesting beach in Florida has someone patrolling, whether it be Patrol volunteers or lifeguards.

The Turtle Patrol was becoming a genuine force in the environmental community, hosting beach clean-ups and rallying for governmental change on top of already protecting nests and transporting sick turtles to rehab clinics. When Beth Libert took the leader position in Volusia in 1992, and then inherited the Flagler permit in 1997 from Fran Guarnieri, who ran the area for years before becoming ill and retiring, the organization was already beginning to grow and mature. It was her higher level of organization and more thorough and formalized training standards (new volunteers usually patrol with a veteran[s] for at least a whole year on average before going it alone), however, that expedited the process, turning a modest club of do-gooders into a full-fledged activist group. And soon, a hierarchy began to form under Libert. Officials were never voted into their positions, but rather they fell into them by experience and general level of involvement. Lori Ottlein became something of a right hand in the organization. Dan and Ramona Rutkowski became supervisors. These veterans of the group were the ones to bring out the newbies to show them the ropes. Once new members were assimilated and ready to patrol alone, they could potentially bring out their own trainees, although some may never be ready for that responsibility.

Before a first training period, newbies are briefed and told to bring water, a snack, sun screen, comfortable clothes. They've been to orientation meetings and given decent introductions, but surveyors like Teresa and Catherine Ryan still openly admit, before that first day, "We had no idea what we were getting into." And that's because each day is different—some days you might find

The Shirley Reynolds Story

Beach driving became an increasingly heated issue along the Florida coast, and Shirley Reynolds, a sort of self-instated champion of the movement, was, in simplest terms, an environmentalist with time and money. In no way officially tied to the Turtle Patrol, the group knew that it had to be strategic in dealing with her and her actions, which was sometimes slippery terrain. In outwardly supporting Reynolds' cause, the Patrol might be viewed by many in the public as the "reason" they would no longer be allowed to drive on the beach, and resent them for it. Public support of Shirley Reynolds had the potential to turn the Patrol into a culprit, an easy-to-spot scapegoat in the town. What the organization needed, though, both politically and personally, was to remain neutral, to be viewed as a friendly face helping the beach, not a bureaucratic one fighting to take it away.

Reynolds sued Volusia County in 1995, and in 1996 a Federal judge ruled in her and the turtles' (who were also listed plaintiffs on the case) favor. The area was subsequently forced to apply for an Incidental Take Permit and develop a Habitat Conservation Plan for sea turtle protection (the first in the state of Florida). Mitigation to keep Volusia County's beach driving privilege included banning night time driving, removing vehicles completely from the nine miles of the densest nesting beaches, implementing a strict lighting ordinance and, finally, building the Marine Science Center of Ponce Inlet.

In 2004, Reynolds turned her attention and took on Flagler County, where beach driving was allowed and unregulated.

It was all a catch-twenty-two for the Turtle Patrol. On the one hand, they backed Reynolds' campaign, but on the other, they couldn't do it publicly. Long-time beach surveyor Lori Ottlein was even asked by Reynolds to be the Flagler signatory on the suit, but after discussions with Beth Libert, Ottlein had to decline. The scenario was this interesting politics game, with many residents growing to hate the turtles for the very prospect that they might be the cause of the halt on beach driving, and the Patrol being careful not to get caught in that crossfire of anger and opposition.

Eventually, Reynolds' case would go to court and the county would ban beach driving in Flagler so as to avoid making room in an already tight budget to establish a Habitat Conservation Plan, as Volusia did. This decision, however, had unanticipated repercussions for the Patrol, which are discussed in detail later.

nothing, others a nesting leatherback, others still (like the Ryan's first day on Patrol) a dead alligator in the surf. During training, new members are taught how to handle themselves in any situation: encountering a turtle on the beach, how to handle eggs on relocation, proper procedures in recording turtle sightings or nests, and basically everything about the day-to-day of beach monitoring. What hardly can be taught, though, is how to differentiate between the six Florida-native turtles and each of their elusive tracks. This is a talent that comes only through time spent on the job, although sometimes handouts can be helpful (see example on facing page).

A DAY IN THE LIFE

Patrolling Duties

The first rule of Turtle Patrol is timeliness. As Peach Hench so aptly puts it, "You get up early, and if you don't want to get up early—don't bother [volunteering]."

Morning patrols usually start on May 1st and as early as 6:00 A.M., lasting as long as eight hours, depending on how many nests are found that require documentation. In a lot of ways, the entire survey process is a race against the clock, considering that Patrollers under no circumstance will use lights on the beach to survey before sunrise, and that 9:00 A.M. is the cut-off for nests dug under the high-tide line to be safely moved for relocation. Let Danny Rutkowski paint a picture for you:

"So there you are. You're out on the beach. You look down the beach and you see six more crawls in front of you and it's [already]

SPECIES IDENTIFICATION

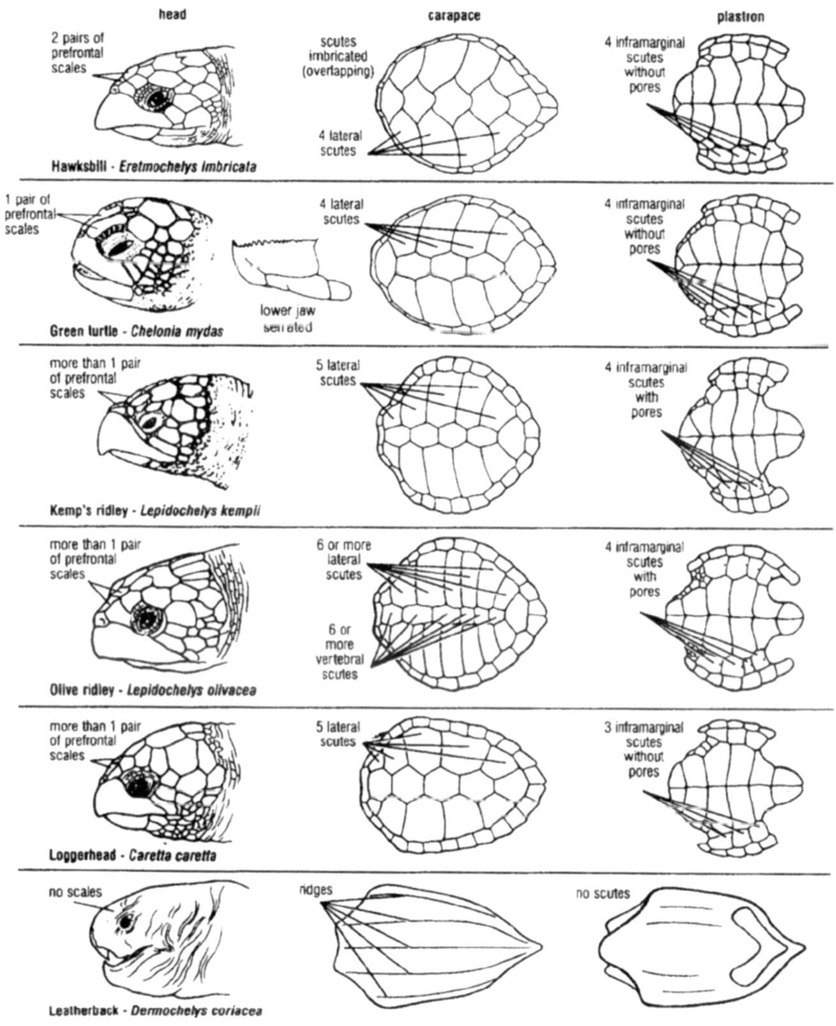

8:00 A.M. You gotta get busy, because it takes about forty-five minutes to move one."

Sometimes being punctual can be especially tough when beachgoers, unaware of the Patrol's fast-approaching deadline, stop surveyors to chat. "Sea Turtle 101" is what they're looking for, says Dale

Hench, "but we have to guard against that." In these busy mornings, Patrollers rush to move all of the nests, sometimes even call for back-up if they don't think they will be able to reach all of the ones in front of them by nine. So during the nesting season, every member of the Turtle Patrol is almost constantly on call and must be ready to fly out of bed and head out to the beach to dig a nest whenever they're needed, a responsibility no Patroller takes lightly.

Whenever paged on their days off, surveyors respond without hesitation, says Catherine Ryan—"because once you're on patrol, you *want* it, you *love* it." She recounted one particular morning where she and her mother found nine nests, six of which were in need of relocation, and how all she had to do was get on the horn and start assigning areas: "Come to 13th Street South—I need you to relocate this nest," she called one volunteer. "Okay, I need

WHAT SPECIES MADE THE CRAWL?

On the following pages are examples of handouts on the different patterns seas turtles make in the sand on their journeys across the beach.

A. Track appearance as in Figure 1, loggerhead track: tracks from a sea turtle with alternating limb movement, no tail drag mark, and track width thirty to forty-five inches.

B. Track appearance as in Figure 1, green turtle track: tracks from a sea turtle with simultaneous limb movement, a center drag mark from the tail, and track width forty-two to fifty-three inches.

C. Track appearance as in Figure 1, leatherback tracks: tracks from a sea turtle with simultaneous limb movement, a center drag mark from the tail, and track width seventy-one to eighty-four inches; track path often circling or sinusoidal (S-shaped).

Note: Flipper injuries to turtles may influence track sign. Characteristics of the nest should be used in conjunction with track characteristics to identify species.

you at 15th Street North," to another. And they showed right up, even brought their own coolers and equipment. "When you're going down the beach on the ATV and it's windy and rainy . . . it kind of sucks," Catherine's sister Teresa admits. "[But] all of us, we just *love* turtles. We may not necessarily love being out in the hot sun or being in the sand, but . . . we do it for the turtles."

The thinking behind the 9:00 A.M. cut-off is that turtle nests are usually laid some time at night after 9:00 P.M., so by 9:00 A.M. the

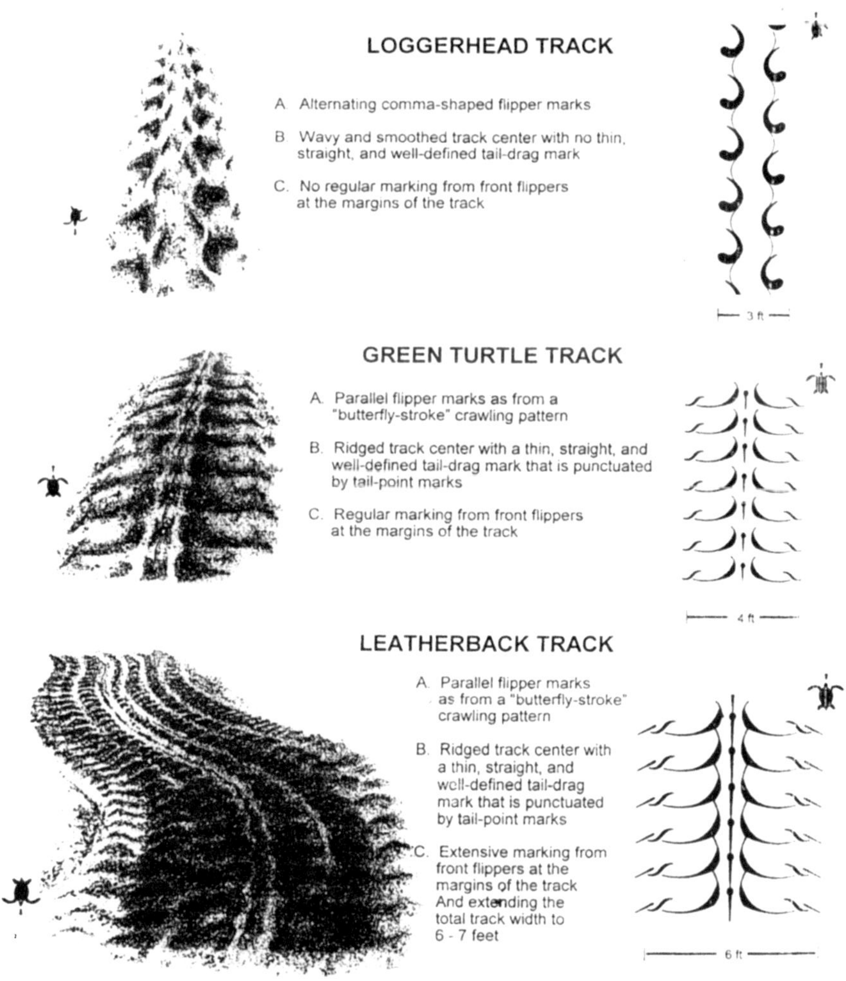

Figure 1. Characteristics of sea turtle tracks found on Florida beaches.

Figure 2. Characteristics of **loggerhead crawls** indicating either that the turtle had previously nested (left a nest) or had abandoned its nesting attempt (left a "false crawl").

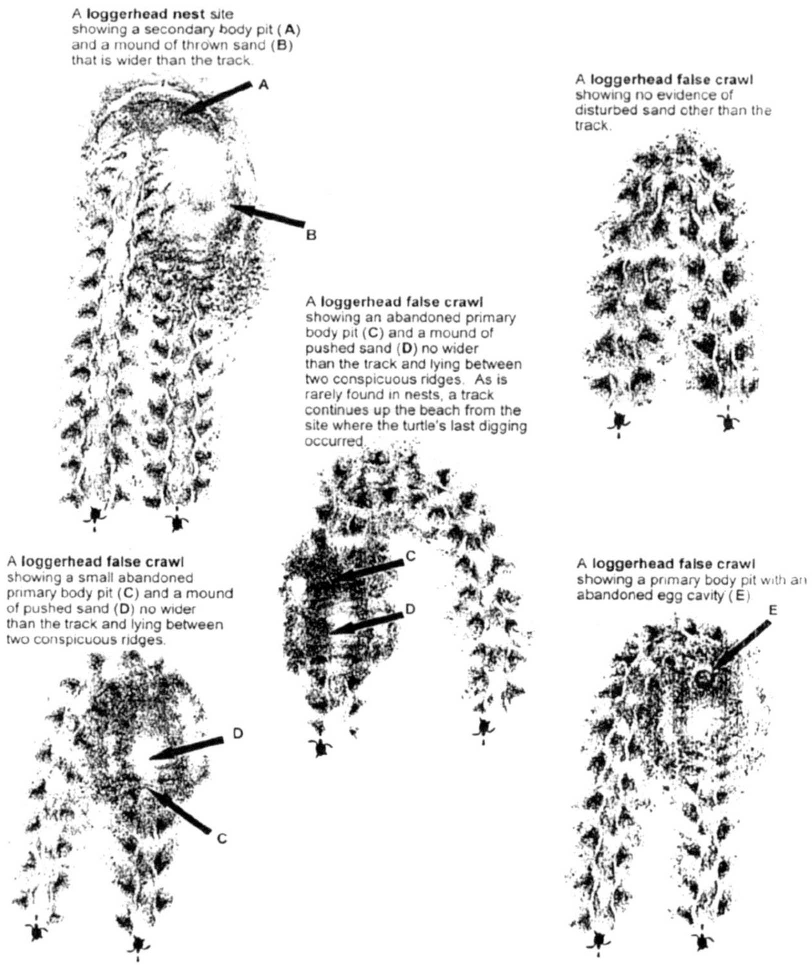

embryo and membrane will have had twelve hours to join. If disrupted after their merger, the two are at great risk of detachment, in essence sterilizing the egg. So great care must be taken when digging eggs out one at a time with gloved hands, and then reversing the process and placing them elsewhere, mimicking the initial chamber's contours and depths, and last, hand-packing the sand at the surface in an attempt to match the first nest's pressure from above. Even the angle of the egg inside the chamber must be repli-

Figure 3. Characteristics of **green turtle crawls** indicating either that the turtle had previously nested (left a nest) or had abandoned its nesting attempt (left a "false crawl").

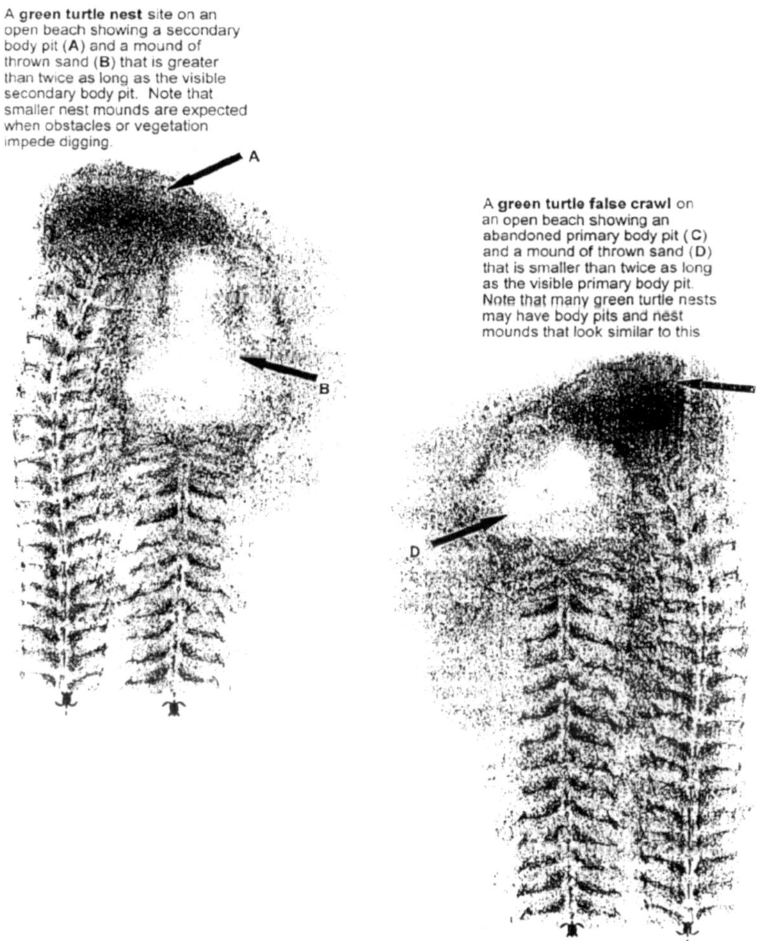

cated by Patrollers moving them. The process is tedious and tiring but necessary in order to ensure successful emergences.

After rising and shining well before first light, surveyors first stop at the official Turtle Patrol shed, double-check that all of their supplies and forms are adequately stocked, mount their ATVs and set off at a blazing 10 mph for the beach, usually fifteen to twenty minutes (tide-dependent) before "civil dawn." During these short trips from the shed to the beach, most Patrollers hope for low-tide,

Lori Ottlein, Teresa Ryan, Dan Rutkowski, and others digging a nest for evaluation.

Catherine: "It's very physically demanding, digging a nest for thirty minutes, an hour, an hour and a half some days. . . . And it's hot, and tiring, and you have to truly love it and have the endurance and—"

Barbara: "Like sand."

Catherine: "Yeah, I mean, you're covered head to toe in sand and you have to *want* it. It's pretty tough."

Teresa: "You find sand in places you didn't know sand could get."

—BARBARA, CATHERINE, AND TERESA RYAN,
SEVEN-YEAR TP MEMBERS

because of how difficult it can sometimes be navigating the beach, especially those areas fortified with rock revetments, when the tide is high and pushing against the dunes. ATVs are used for every survey unless the weather is especially bad. In these cases, if it's toward the end of the season, surveyors in Volusia County will traverse their six-mile-long appointed zones in cars, ordinarily, slowly and carefully rolling down the beach with their eyes peeled. But ATVs are preferred.

"My favorite part of patrol," says Catherine Ryan, "is when we take the ATV down the hill [toward the beach]. That's my favorite part. Every single time I take the ATV out, I stop on the hill—because the sun's not up yet. That's your first real blast of that ocean, that salt air. And I always sit there just for a second. And I take a breath. And it's like, (whispers) '. . . wow . . .' "

Once en route, basically what surveyors are doing is looking for tracks, tracks that lead to nests, tracks from emergences, false crawl tracks. In the beginning of the season and for the first one-and-a-half to two months, marking nests stand as the volunteers' primary duty. They drive near the water where the ground is still damp so as not to leave tracks in the soft sand, and look for turtle prints. On sighting a nest, each one is marked with stakes and ribbon (four stakes around the nest, one in the dune above the nest as a security measure in case the others get washed or blown away); the species that made the crawl is ID'd; and the proper paperwork is filled out. Then, it's basically maintaining the status-quo until the hatching starts. During that interim, surveyors check each nest already laid, making sure none have been disturbed or vandalized. And at the end of each day, ATVs must be brought back to the shed and ritualistically cleaned. These vehicles shouldn't be driven through the surf, but on the beach corrosion is unavoidable. Volunteers drive these four-wheelers hard through damp sand and salty air every morning, so hosing them off after patrol is necessary in order to prolong the life of these machines for as long as possible. Even with the daily cleanings, though, Teresa Ryan estimates that each ATV the Patrol acquires probably doesn't last any longer than three or four years. The beach takes its toll. So it goes.

Three days after the first emergence, each nest is evaluated, or "cleaned," by surveyors counting to see how many eggs were laid, how many emerged and how many are left unbroken inside. Any eggs still intact are placed back underneath a more shallow area of sand to continue incubating in hopes that they are just late bloomers and not infertile. Nest cleaning is an aspect of the job that might be easily overlooked, but it stands as one of the primary tasks the Patrol conducts in collecting its yearly nesting numbers. Unless a nest requires relocation, Patrol members will have no clue how many eggs were laid in it until emergence—could be 50, could be 179, like the nest dug by Lori Ottlein a few years back, the largest ever dug by a Patroller in the Flagler area. The number of hatchlings is recorded by counting the broken egg shells left in the chamber. Surveyors do this as best as possible, counting a shell only if more than fifty percent of it remains. It's a semi-subjective count, and though Patrollers might know for certain that there are enough fragments to account for another two or three eggs, they follow state guidelines, always slightly undercounting the clutch for the sake of accuracy. The evaluation is made more complicated with further emergences from the same nest. These, too, are left as judgment calls for volunteers. When tracks are first recorded from a nest, it has to be decided whether enough of them are present to warrant labeling the hatching as a "good, strong first emergence." Sometimes this first breach can be only one, eight or ten hatchling tracks; other times it can be dozens. Then, every day after that, the nest is checked for further movement to be recorded.

This monitoring process is more or less generally the same for beach surveyors everywhere, except in certain locations, like Rancho Nuevo, Mexico, where arribadas occur. On these exceptionally dense nesting areas, evaluators may record nest location but not stake each individual nest—there are just too many of them. They also might not clean out each nest, either, for lack of manpower, even though evaluating nest success is standard protocol in the Volusia/Flagler regions.

Each day on the job will vary for Turtle Patrollers, as well, as far as how long surveys can last and what tasks are completed. On the beach at 6:00 A.M., Patrollers can be out as late as 2:00 P.M. if many nests require relocation and marking, or if many strandings or washbacks are found. All of these discoveries require that subsequent documentation be completed and filed, as do track and false crawl sightings. On the flip side, sometimes surveys are a breeze, over and done with in only two hours if no special circumstances arise. Volunteers are out every morning but there's a rotation in place, with some hitting the sand and covering their six to twelve miles only once a week, and others as many as three times or more.

These aspects of the monitoring process are mainly relegated to the summer months but, in actuality, the Turtle Patrol works year-round. Although there are no nests to mark, move or excavate in the winter or spring when turtles haven't yet returned to Florida beaches, there are still strandings to deal with and transport to rehabilitation centers—and because not every Patroller is trained to handle strandings, a select few are burdened with double the work. There are years, like the year this book was written, when the weather is so severe that it brings thousands of cold-shocked turtles onto the shore, some dead, some lethargic and barely hanging on, and all needing to be handled and documented accordingly. There is the transporting of turtles to marine centers, a task that is often shared between Patrollers of different counties. (A loggerhead might be found in a county two or more hours away from the Marine Science Center, driven to Flagler for one of the local Patrollers to share the burden and take it the rest of the way to Ponce Inlet). And there is also genetics research conducted in the off-season, wherein samples are extracted from a dead stranding, stored and taken to marine labs for use in DNA records that will help in calculating turtle population diversity.

These muscle samples are studied by the Fish & Wildlife Commission for various reasons. The two-year sampling of dead hatchlings is part of a research project being conducted by a graduate student at the University of Georgia to determine genetically diverse populations. Samples from nests were salvaged by all volunteers. Teresa Ryan in Flagler County and Beth Libert in Volusia extract this muscle tissue from the flippers of deceased turtles and preserve them in vials.

None of this would be possible, of course, without the proper training, and the best training is hands-on. When one signs up for the Patrol, they're given handouts and orientations, but the real learning doesn't begin until their first day on the beach. Some people get unlucky, though, and never see a single track for weeks. In cases like these, no intensive OJT (on-the-job training) occurs, and so it takes trainees longer to get assimilated to the routine of 1) spotting tracks, 2) identifying what species made them, 3) identifying if they are to a nest or from a false crawl, 4) identifying if a given nest is in danger of being washed over, and 5) marking or

CHILL OUT

When dead strandings are found on the beach, a small piece of muscle is extracted from the turtle for uses in constructing family trees. For this kind of sample, it doesn't matter what stage of decomposition the carcass is in–only that a marble or so-sized chunk of muscle be cut from the creature (this is usually an entire fin from hatchlings, or a piece of fin from adults). Once extracted, another incision is made into the muscle itself to ensure that the buffer solution penetrates the tissue. Then the sample is placed into a jar or test tube for analysis.

"But it gets better," said Barbara Ryan after her daughter Teresa explained that part of the research process. See, the samples have to be stored and preserved if they are not brought immediately to a lab, which are usually hours out of the way and sometimes hard to fit into a day's schedule. "So Teresa, being Teresa," Barbara said, "would bring them home–and put them in our refrigerator." And this is where they stayed, whole lines of them, all along the door. Barbara smirked and eyeballed her daughter as she said all this. Then she started laughing. "[You go to get ketchup] and it's like, 'Oh my God, get these out of my refrigerator!"

Why Barbara really told us all of this, though, wasn't only to embarrass her daughter (even though, having seen their dynamic in action, it's safe to assume that probably didn't hurt), but to illustrate just how involved Turtle Patrol can be. "It's not just, 'Oh, let's go for a ride on the beach'," she said. And for Teresa, she does this job because she knows that she's one of the few who doesn't mind it. She's taking one for the team, contributing to the group effort the Patrol already boasts in their making of schedules and response time to off-day pages. She's doing the dirty work so that her friends don't have to, saving them the grief. Look at it that way and the job actually seems kind of sweet, not so . . . you know, morbid and repulsive.

relocating a nest. All volunteers are also required to attend yearly state training workshops.

Still, even of the trainees who get lucky enough to jump into the work headlong, not many will make it past the basic period and go on to be lifetime members like those showcased throughout these pages. According to Danny Rutkowski, in fact, the percentage of converts is astonishingly tiny.

"Come back the next year?" he asked. "One percent. You sort them out. They get sorted out. Once it heats up." Of that estimated one percent, some may never even be ready to handle a solo patrol or take on a trainee. Either they never acquire the skills to do so or they simply see patrolling as a more casual exercise. Others can even be fired if they overlook too many nests or misuse the ATVs. They can be fired from their volunteer jobs.

"Quite a few people come out on patrol once, twice, three times, then they'll [stop]," says Catherine Ryan. "If you make it through July—you're a lifer. But you have to make it through July."

Put yourself in a trainee's shoes and it's hard to imagine it not being disheartening, lumbering through a storm-torn beach in the thick heat of a Florida summer, being drenched, alone, thirsty and tired. The only occasion when volunteers *don't* patrol is if the sky is on fire, exploding in streaks of hot static across the cloudbursts. But for some, these harsher conditions are small prices to pay. "[It's] sort of like the mailman thing," says Dale Hench plainly. "It's your day."

For Dale and Peach Hench, the good of patrolling has always outweighed the bad, even from their very first solo nest cleaning when, after all of their excitement and anticipation, the only nest they had was drowned and rotting. "It was awful, every one dead. It was nothing but dead turtles, and it was horrible," Peach said. "And we still had to evaluate [it], had to dig them all out, and it was bad. It was the worst."

"Really, there's nothing that smells worse than a rotten turtle nest," Dale agreed, an assertion no Patroller would challenge. But surprisingly, this first encounter left the Henches not the least bit discouraged. They had seen nest evaluations before, knew that it was always a toss-up whether the experience would be fulfilling or solemn: whether they would be carefully extracting broken shells and looking for happy stragglers, or if they would be holding their breath, reaching for dozens of decaying corpses to be pulled onto the beach for a death count. On later patrols, they would evaluate nests that were one hundred percent hatched, and those would be the redeemers. Those are the ones that kept them coming back.

Turtle Patrolling is strange and unforgiving work, and sometimes, surveyors admit, all they want is to get off the beach. The long, early hours, the scorching or rainy weather, the ants and flies and sand spurs—it's definitely not for everyone. This natural filtration, however, although steep, leaves the team with only the most dedicated of volunteers, those willing to leap out of bed in the middle of the night to tend to a nest, those who don't mind lying face-down in the sand, in a storm, reaching in earth up to their shoulders to excavate eggs to safer ground. These members seem somehow to be built for this kind of work, the severe elements doing nothing but building a certain kind of character within them, making them look forward to the next success only that much more. These are the people, like the Henches, like the Rutkowskis, who can look you straight in the eye when you ask them what their least favorite part of the job is, and confidently, unflinchingly say, "Nothing."

"You're down there 6:30 in the morning and you may be down there until 2:00 in the afternoon," Rutkowski says. "It's ninety-five degrees and you're in the sand, and you're in the bugs, and you're in the heat. And a lot of people will drop out because of that. It's just that the reward's not enough."

The "reward." He says the word plainly without explaining what exactly it is or what it means—but, then, he doesn't have to. It's doubtful he even thinks about it too much, himself—lifestyles are usually that way, not something to be justified. Presumably, the reward is having something tangible to identify oneself with, someplace nice to take family photos, something to talk about. The reward is the whole reason he came in to interview. The reward is why he's here.

Lori Ottlein

Member of the Volusia/
Flagler Turtle Patrol:
2000–Present

"I have sent [in] many pictures and hope that you do not hate me for it!! But I wanted you to see some of the bad, some good, and some stressful sides of Turtle Patrol!! Every one has a great picture of Turtle Patrol but it is not always so pretty. BUT!! The good soooo outweighs the bad!!"

—excerpt from e-mail

Of all the members of the Turtle Patrol that we met with while putting this book together, Lori Ottlien was, without contest, the most eager. She's a "smiler," one of those people you see around town that seems to be happy all the time, twenty-four/seven, their enthusiasm infectious and impossible not to catch. The owner of sixteen water turtles, most of which were rescued or inherited, and a couple of tortoises, Lori laughs a lot and is the type of person to attach a line of exclamation points after everything she types. She brought in a stack of papers and brochures for her interview, e-mailed in dozens of great photos with captions, and told us that she had to restrain herself when one of us asked the day before our meeting if she could send in more. "You are a silly guy to encourage me!!!" she wrote. "You may be sorry!!!"

In no time after her arrival, we were laughing and talking turtles, her session a more casual conversation than a series of questions and answers like all the rest. She was animated, and talked about the dirty diapers people leave on the beach and the annoyed faces they make when she picks them up and tries to return them to sender. She chuckled, and recounted the time she was almost arrested, then reminisced about how nervous she and a few Patrol-friends were before getting their first (of two) turtle tattoos.

No matter what quotes we showcase or how we choose describe her, it just may be impossible to convey on paper Lori's enthusiasm or general level of excitement. Letters are just too stiff.

[excerpt from interview]

Lori: It's all wow. Starts in the beginning of the season. You're like, "Oh wow" about seeing the tracks and how amazing that is, that this 300 pound *monster* came out of the ocean, you know, digs a nest *this* narrow, puts all of her eggs in there. . .And that's amazing. Then the season goes on and you see all these baby tracks, and *that's* amazing. And then you see the little hatchlings coming out and you're like, how? It's as deep as my arm! How do they do that?

. . . So weird how they do the same things and come up and make their nests right where they were born. There will be a nest laid on Sunday and then three or four days later there's another nest and it's like, "Oh my God, *sisters*!" (laughs) You can put so much into this, oh man . . .

How about the worst moments?

Lori: The winters of no turtles, you go through . . . you just need to be out there.

Turtle withdrawal?

Lori: Just like turtle withdrawal. Just like, "I've got to get back in touch!" . . . But, I mean, the saddest thing is to have to go out and do the strandings. I probably wouldn't do it if I hadn't started doing it, because it's very bad. . . . A lot of times it's chopped up with boat propeller marks or things like that. [Sometimes] you find huge leatherbacks, rolling back and forth, back and forth. I mean who knows why it died, but those are the sad ones.

Do you have any final thoughts? Any lasting impressions you want included?

Lori: Just how cool it is. I mean, it's funny because you get the new volunteers and I think almost every time you get them, they go down there and they're excited, and then it just becomes, all of a sudden, they're . . . *absorbed,* you know? [At first they're just], "Huh, that's kind of cool. Oh, I'm glad you're doing that. That's nice." But then it seems once you get into it, you're like, "Oh my God, *turtles*!" And that's it. Everybody gives you turtle stuff, birthdays, you get turtle stuff for Christmas. You just can't pass a turtle object.

Because that's your thing.

Lori: Yep, it's easy to buy presents. I want a turtle, a piece of cake.

Many people talk about it like it's an addiction. Once you get in, you have that first experience, it inspires you, and you're addicted.

Lori: It *is* an addiction. Some people leave and we don't really understand it. I know I couldn't give it up. But there are people who do it for a couple years and never come back. I guess they find something else to do. (shrugs) I don't know.

It's got to have something to do with the lifestyle. It's almost like anything—when you start doing it, you get used to it in a way, and to stop seems weird. "Why would I stop?"

Lori: Missing it, you'd have to be missing it. I mean, it's also you have to really love the beach and want to go to the beach all summer. . . . That's my life. I mean, everybody knows, all my friends, all summer long it's like, "Oh well, you'll see her when turtle season ends. Not before."

Is it a matter of getting close and deciding, "Am I going to get into this that much?" Then once you make that leap, there's no turning back?

Lori: Nope, there's no turning back—at all. Then you start getting tattoos—turtle tattoos, I mean. Last year we went to the permit holders meeting and, [after], three of us went to the tattoo parlor and, all three, we all have the same tattoo.

(laughs) It's like a gang sign!

Lori: (laughs) It's like a gang sign, that's right! I was thinking, "Don't you remember a long time ago all those girls in the Olympics on the swimming team that all got the same tattoo? That's what *we* need to do!" But no, some of them are scared.

I guess you've made a family of it, too, because you're all branding yourselves with this turtle tattoo. You're all part of the same group.

Lori: We all just love the same things—that's what's cool. Like, this year's the first year Teresa [Ryan] came to the meeting and I said, "You've got to come because you're in this huge auditorium and you're going, '. . . that many of us? How can there be so many people that love the turtles?'" Because they have the same mind. . . . Of course, again, winter time you pretty much [separate]. You get to see your other friends.

Your "real world" friends.

Lori: The summer comes, you do nest cleanings. You go out to dinner or go eating before . . . But yeah, all summer long you got this gang. Then in the winter, people go on vacation, people go here, [there]. Nobody can go on vacation during the summer. I mean, how would you? You can't! You'd miss your two days of patrol. You may miss something really good that week.

In one of our final e-mails, Lori closed with the following:

Lori: I really would LOVE for you to go on patrol with me one day during a good time during Turtle season. You just may get the sickness!! Seventy-eight days to go before the 2010 season starts!! WHOO HOO!!

CHAPTER 5

Going the Distance: The Eye of the Turtle

By the early 2000s the Turtle Patrol was pretty well established. The region of Florida's Flagler County had been added to the Volusia routine, and instead of driving the beach in personal pick-up trucks during surveys—which Beth Libert did for years, burning through three of her own vehicles in the interim, using them until the sand and salt took their final toll—surveyors drove county-sponsored clunkers and then ATVs. By this time, the TP was beginning to garner all kinds of attention. Their constant presence on the beach and their success with raising yearly nesting numbers had been enough to make the faces of individual surveyors pretty distinguishable on the street. Fishermen began recognizing them at dawn, nodding their way or asking about sightings and nest counts as they scooted along on their four-wheelers. Newspapers were regularly writing articles

So darn cute!!
"Look out world, here I come!!"

We were to clean this nest which had emerged three days ago and when we started to clean it all the hatchlings just started busting out!

Caption by Lori Ottlein

highlighting their efforts. The term "household name" may seem a little dramatic, but it's what they became. For maybe for the first time ever, people *knew* the Turtle Patrol.

At this point, it would have been easy for the group to grow complacent, rest on their laurels, change their passion for glory. But Beth Libert and the rest of the Patrol's elders, so to speak, knew that the only real way to make the impact they set out to secure was by reaching as many people as possible with their message. Education, they knew, was the key, creating a feeling of public ownership over the animals, ensuring that the people, especially the kids, knew as much as possible about the endangered creatures native to their backyards. Education was the cornerstone of the Turtle Patrol's plan for expansion. Only through it could they create a new norm, a fresh air of respect and protection over the creatures they were fighting to save.

In 1995 the Turtle Patrol published the first issue of its thrice-yearly newsletter, the *Turtle Tracker,** where they informed of the group's regular activities and patrols, explained various issues on which the Patrol needed support, and offered opportunities to interact with sea turtles up close and attend presentations. The newsletter cost only fifteen dollars a year, with all profits, and any other donations received from it going toward funding the Patrol, buying supplies, gas, new (read: used, but new to the Patrol) ATVs, and anything else needed to conduct daily surveys. The Patrol was and is a non-profit organization and, once it grew to the size it reached in the mid-1990s, began heavily relying on public donations to continue its work.

GREENBACKS FOR GREENBACKS

Enduring only through the charity of others, over the years the Turtle Patrol came to realize that the best kinds of donations are not

---

* In 2004, after nine years of publication, the final issue of the *Turtle Tracker* was published. The project was put on hold because of time and sale-related limitations. Beth Libert says she'd like to bring the newsletter back one day—and electronically—but right now it's still up in the air.

1995 *Volume 1* • *Issue 1*

Turtle Tracker

Volusia Sea Turtle Volunteer News

The First Volusia Sea Turtle Patrols

The first organized turtle patrols in Volusia County were started by the late Richard Durrbeck. It all came about when Richard's daughters wanted to do a science project on sea turtles.

In order to find out more information about sea turtles, Richard, his wife Joyce and their two daughters, Sharon and Stacey visited *The Turtle Lady*, in Ponte Vedra Beach. There they saw a release of Loggerhead hatchlings and quickly became hooked on turtles, as so many of us have. The family also visited the House of Refuge on Hutchinson Island in Stuart. This is a museum of the sea and at that time was a holding facility for injured sea turtles and a head start hatchling program.

In the early 1980s, the Durrbecks became so interested in sea turtle preservation that they got Ross Witham, who was the sea turtle coordinator for DNR *(Department of Natural Resources)* at that time, to help them apply for a permit to do nesting surveys and relocate nests.

Since there was no one else doing sea turtle patrols in Volusia County at the time, the Durrbecks became responsible for the area from Ponce Inlet north to the Flagler County line. There were no local ordinances protecting sea turtles then, so all nests had to be moved. Beach front lighting and vehicular traffic was not regulated so the sea turtles stood very little chance of survival, being left where they were laid.

The answer to increase the chances of survival seemed to be artificial incubation in styrofoam coolers in the garage. This later proved to be an unhealthy practice for various reasons and is not permitted by FDEP *(Florida Department of Environmental Protection)* anymore.

A couple of years into this project Richard realized how difficult it was to adequately cover so much beach. He then recruited Beth Libert to cover from the Main Street Pier south to the Ponce Inlet jetty. Shortly after that, in 1984, Marye Marshall and her husband Frank assumed responsibility for New Smyrna Beach, from Ponce Inlet to Canaveral National Seashore, creating Volusia Sea Turtle Society. All of Volusia County was being covered by volunteers.

When Richard passed away in 1992, Beth Libert started the Volusia Turtle Patrol, Inc. and took over the permit for the 25 miles north of Ponce Inlet.

"We miss you Richard!"

Written by Beth Libert, President, Volusia Turtle Patrol. Look for more information on Volusia Turtle Patrol and Volusia Sea Turtle Society in future newsletters.

What it involves

Every morning the dedicated volunteers survey the beach between the north Peninsula State Recreation Area and Canaveral National Seashore. *(Park rangers and Park volunteers cover park areas.)* Turtle volunteers are trained and then permitted by the Florida Department of Environmental Protection. They follow strict guidelines and regulations for conducting surveys, marking, monitoring, or relocating nests. Every year the guidelines are updated with the latest information research has documented.

strictly monetary. If the Patrol can get the person or organization donating involved more than just financially, their contributions are working two-fold, essentially doubling the project's effect on the community.

Monetary donations are always welcome, of course, but, in a way, they're more impersonal, detached. With many of the group's

recent fundraising initiatives, they've tried with great success to mobilize physical involvement, earning in the process a heightened sense of awareness in the community by imparting to the public what the group has known forever: turtle conservation isn't just rewarding—it can also be fun.

"We'll name him . . . Bulbasaur Smith": Nest Adoption

Their most successful campaign to date, the Turtle Patrol's Adopt-A-Nest Program puts responsibility for the turtles straight into the people's hands. When a family, company, or, best of all, a school class throws down thirty-five dollars to adopt a nest, it becomes "theirs," and they can choose an In-Name-Only adoption or a Hands-On one. If Hands-On is chosen, on top of receiving the highly prestigious Turtle Patrol adoption certificate, a bumper sticker and a fashionable sea turtle shirt, adopters are also encouraged to steward their nest by routinely monitoring the area, shooing away predators, smoothing out

Lincoln Hunt stands next to one of the many nests his family adopts yearly.

holes and tire tracks that may be blocking a hatchling's course toward the ocean, and various other things. The main attraction of the program, though, is the nest's emergence. Unlike nests offered by many other conservation organizations, the nests the Turtle Patrol puts up for adoption are real nests laid by wild sea turtles in the most current nesting season. Nest adopters will be called by a Patrol member days in advance of a predicted emergence and invited to come out to the beach to witness the event, to watch the nest they helped save turn into an overflow of real, live turtles and crawl into the sea. When a school class adopts a nest, teachers are encouraged to schedule field trips or after-school meetings so that the students can see in person this most intimate of natural phenomena. As unpredictable as emergences can be, though, and the fact that they usually occur at night or dusk, the nest "cleaning" generally ends up being the event worthy of a class trip. A presentation will be given about the nest, where the turtle deposited its eggs, how the hatchlings made it out. And sometimes, there are still hatchlings inside for the kids to see. Dale Hench is usually the one to conduct these presentations in Flagler County. He's usually the face.

"We love to have him," Lori Ottlein said, "because we can say all the same stuff, but as soon as I start with 'And then the mother turtle comes up on the beach . . .' [Pretends to start crying] 'Oh my God, it's so great!' [But] he's got this great booming voice. He can tell you the facts without getting all emotional."

Before the Adopt-a-Nest program was instated, Patrollers would invite anyone they saw nearby to come to nest cleanings to either watch or help dig. But today, when civilian digging is no longer allowed and nearly every nest on the beach is accounted for, they try to make the experience smaller and a bit more personal for those who've invested. Many of the nests claimed throughout the year are bought as gifts, and some surveyors don't feel right about inviting the whole neighborhood to share in an event that one particular party of people reserved months earlier to share with their families.

Other times, though, evaluations are a huge deal, drawing a couple hundred people to the beach. These are the instances where Patrollers like the Ryans will bring along friends for help in education and "crowd control." "Once you've got the babies and you're digging them out of the nest and they have to walk to the water," Barbara Ryan said, "everybody wants to get pictures; everybody wants to see them; everybody wants to touch them." But they can't let the people touch them. It's integral to each hatchling's beach imprint that they crawl from their nest to the water on their own and without interruption. Disruption can result in disorientations.

Barbara's daughter Catherine agrees, "Sometimes you have to be pretty blunt but, you know, we're here *for* the turtles, not for *you* to see the turtles. Just like when we're opening up a nest, people will go stand in the dunes. We're like, 'Guys, you got to come out of the dune.' You know, we are here to conserve the beach and conserve the turtles—not ruin it in the process." So it's a fine line to walk, between welcoming public interaction and maintaining a natural environment for the hatchlings.

Depending on temperature and rainfall, hatchling dates will vary from forty-six to seventy days on Florida's peninsula. Even though hatchlings can emerge at any time, dusk is generally when they sense the cooler evening temperature and attempt their first crawl. Adoptive nest parents are told all of this information, as well as clues to watch out for that will signal that an emergence is near. If the sand at the top of a nest is cracking or caving in, it means that a nest may emerge that night or very soon. It's what Danny Rutkowski refers to as the nest turning into a "pancake," because of the way it appears to bubble and then indent into a circle shape. These holes are not the same as ghost crab holes (which are small and round and usually occur in or around a nest, the crabs generally doing minimal damage to the eggs inside). Nest "pancake" holes are usually larger, occurring because of egg and hatchling movement

underneath the sand. This movement of weight shifts the structure of the nest a bit, resulting in crevices and gaps that collapse down onto the eggs. Signs of this collapse serve as a dead giveaway that a clutch is soon to emerge.

If a new owner opts for a "Hands-On" adoption, partial responsibility for a nest is assumed. The adopter will become something of a nest-sitter, checking from time to time to ensure that conditions are on the "up and up." Things Hands-On owners should look for when checking their nest include:

1. Broken ribbons or stakes to be reported to the TP
2. Tire ruts in front of the nest (greater than 3 feet wide and 1 inch deep)
3. Street and residential lights visible from the beach
4. Deep holes near the nest
5. Signs of hatchling emergence (tracks or sunken nest cavity)
6. Predator tracks, such as raccoons or dogs
7. Obstacles near the nest that may trap or hinder a first crawl (trash, seaweed, chairs, tents, etc.)

Things to do to ensure adopted hatchlings make it safely to the sea:

1. Smooth out foot and tire tracks and fill in any holes, including ghost crab holes
2. DO NOT try to help hatchlings out of the nest—digging a nest is against the law
3. DO NOT let people handle hatchlings—also against the law
4. If it is dusk or darker, make sure that nobody takes flash photography
5. Do not use flashlights

On emergence, hatchlings tend to fan out as they make their way toward the ocean—this is completely normal. The only time nest owners should ever interfere with a first crawl is if a hatchling is heading directly parallel to the water or toward a dune. In these cases, hatchlings should be picked up and moved close to the water, but not into it. Hatchlings must have the experience of their first crawl to build the strength to take on the tide, also to form their natal beach imprint. If hatchlings continue to crawl in wrong directions, contact the Turtle Patrol immediately for assistance.

Above all else, the Patrol stresses that their Adopt-A-Nest campaigns be fun without disrupting the fragile course of "Mother Nature" unless absolutely necessary. It is illegal to "harm, harass or possess sea turtles or their eggs" and the guidelines in place regulating these laws are strict. Also, sea turtles can carry salmonella, which can be fatal to humans. Although the threat of bacterial infection is light, if you have to touch a hatchling, be sure to wash your hands thoroughly afterwards.

The primary handle of the Adopt-A-Nest program, what it's really all about, is getting people out there, on the beach, close to the turtles and able to interact with and relate to them. It's about creating a bond, a bond most readily glimpsed in the faces of thirty second-graders, watching anxiously as the nest they helped tend finally emerges. Their mouths agape and their eyes glazed in wonder, you can see it, right there, as they watch a hundred fragile little bodies push themselves from underground and crawl toward the sea.

"These little ones are our future," Danny Rutkowski says of the classes of students who view Patrol releases and cleanings. "We will have ten to one hundred kids. When we clean the nest, we do a verbal presentation to them: 'Here's what came up, here's where she laid her nest, here's where she crawled out.' . . . Usually there are babies in there so the kids get to see the little ones. That's when we hand out the information."

"When they see it, it just makes it work," Lori Ottlein adds. "It becomes real. Now they're going to stop the car when they see the gopher tortoise on the side of the road. And they're going to save it, rather than just [she throws her hands in the air] *whatever.*"

These Colors Don't Run: Turtle Patrol Paraphernalia

Although not as interactive as nest adoptions, the Turtle Patrol gains a sizable portion of its yearly revenue from tee-shirt sales. Originally designed by award-winning artist Mary Anne James, tank tops, kids shirts, even tote bags are made and distributed to the public to raise Patrol brand recognition.

Turtle license plates are another big novelty item for the Patrol. Thanks to the Sea Turtle License Plate Grant Program, the Patrol once received enough to print informative brochures to be handed out up and down the coast, to individuals or any business that would keep them and make them available to its frequenters. The brochures are fairly simple: pictures here and there of tracks and hatchlings, information on the turtle disease fibropapillomatosis and how to handle stranded or injured sea turtles, contact information, the commercial fishing threat. They serve as sea turtle starter-packs, in a way, an entry-level look at a resource too many Floridians simply know too little about. Like the tee-shirts, the brochures are just an introduction, something constant in the community to keep residents privy to the issues at hand and remind them to take them seriously. Since 2001, the Sea Turtle License Plate Grant Program has awarded more than two and half million dollars to sea turtle conservation throughout Florida.

So far, the idea seems to be effective. Coupled with the volunteers' daily presence on the beach and their participation in 4th of July and Christmas parades, many members of the Turtle Patrol feel that most residents respect what they do and, in turn, respect the nests. Dale and Peach Hench, as examples, claim that many on the beach know the two by name at this point; others, those they've never met, often stop them to ask about their grind. Several have even elected themselves as honorary members of the Patrol's ranks, not necessarily attending cleanings or able to go on patrol because of time, age* or disability, but walking the beach every day in search for fresh tracks, stopping surveyors like Dale and Peach to tell them that they've been keeping an eye on particular nests for them when they're not around. One even caught a poacher in the act and managed to save the clutch that was being stolen.

"It was a friend of the turtles and the Turtle Patrol who walked the beach every day and he called the police," said Dale. The man was spotted stealing from a nest in broad daylight, arrested, and eventually deported for his breaking of federal endangered species laws.

Another group of honoraries confronted the Ryans one morning when they were out without an ATV. "What are you doing? Why are you messing with our nest?" they asked them, and the Ryans had to turn around, point out their Patrol shirts and explain who they were. It was a shock, to be dressed down by strangers like that, strangers who have made it their unofficial duty to play bodyguard to the beach's nests. "But it's nice," they said of these neighborhood nest bouncers. "And you always thank them."

That any of this assistance or notoriety is inextricably tied to the marketing of tee-shirts or license plates is unlikely, but it is tied to the same idea of name recognition. When people know who you are and understand your struggle, they're much less likely to undermine it. By becoming such a presence in the community, the Patrol has inadvertently created its own network of helpers and inside men, keeping their eyes and ears on areas of the beach when the Patrol is not around. In a lot of ways, they've made a Turtle Patrol out of much of their surrounding area. They've made a community.

Lori Ottlein leads Turtle Patrol in Christmas Parade.

* Ruth Schwarz, Turtle Patrol's eldest member, has never let petty things like age get in the way of her volunteer work. At eighty-nine years young, she has been working the beach beat for the past three years.

Backyard Brouhaha: Turtle Fest!

"Turtle Fest is great because it brings the turtle-loving community together and promotes awareness and education."

—TURTLE PATROL PRESIDENT, BETH LIBERT

Latest of all Patrol fundraising ingenuities, Turtle Fest is the biggest turtle-themed happening this side of the Atlantic. In only its third year, the event has grown incrementally since its inception in 2008, when many in the Patrol were expecting a first-year bust but saw an outpouring. It's a celebration, a family event with live music, food, environmental exhibits, face painting for kids, raffle prizes. It's a town-wide get-together where Patrollers bring their pet turtles and tortoises, some of which have grown to over sixty pounds, to show off for display. There is even the possible release of a rehabilitated turtle back into the ocean scheduled on the agenda. The event brings turtles and people together without the feeling of commitment that might accompany going to a meeting or leafing through a brochure. It makes the information available but presents it in warm living color behind carnival face paint and a live band, then lets natural curiosity take its course. It's the same idea as when teachers bring "class pets" in for their students to gawk at and point at and eventually love.

Without fail, every member of the Turtle Patrol we spoke to while putting this book together brought up the Turtle Fest with a certain excited gleam in their eye. The first Fest was good, a lot of them told us; the second was better; and the third they're hoping will just explode. These members know that their new creation has great potential to become the Patrol's most important day of the year, and it's a prospect they embrace.

The goal of the Turtle Fest, in addition to public education and Patrol fundraising, is to capture a weekend representation of the "Flagler experience." Life is different for those on the coast, and the Fest celebrates that fact, celebrates the wonder sprawling in every direction to a vanishing point right in the town's backyard. It's the southern equivalent of a neighborhood block party, where the town meets in a beach-side square to picnic, barbeque and relax with their families. There are even tortoise races—a favorite among the children—where tortoises are placed in walled rectangular paths lined with hibiscus leaves and cheered as they make their way from one to the other, and finally, to the finish line.

"Kids everywhere," Lori Ottlein said, laughing, "'Go Pokey!' Some of us become babysitters because the kids [With her hands, she mimes children peaking over a barrier]—for like three hours! . . . They just sit there and play with the turtles for hours."

The first year, though, the tortoise races became a real hot-button issue when the Humane Society witnessed one and aired its chagrin. Representatives didn't approve of the contests, Ottlein said, and the Patrol didn't understand why. It wasn't like they were pushing the turtles along, pitting one against the other and chanting, throwing money into the ring on whether it would be Tank or Streaker to crawl out victorious. This wasn't the turtle Coliseum. They simply let the animals go, taped names onto their shells and watched them eat hibiscus. The Patrol went out and did some investigative journalism, checked out similar turtle-centered soirées and, as they presumed, most of them organized some form of slow-motion footrace. That affirmation was enough to convince them to keep the events around.

"Okay," Ottlein thought, "we're not that bad; we're not doing anything. [And] the kids really do like it."

* * *

Thanks to these various fundraisers and the donations the Patrol has consistently received, the group has been able to continue its work fairly unburdened by financial stressors for roughly thirty years, a feat that would have been impossible without the help of its community. This largely unanimous town-wide support is something members of the Patrol witness every day, and every day it still manages to surprise them.

"One time," Peach Hench reminisced, "this is probably four or five years ago, early on, we were on patrol and this woman came up to us and handed us a check for five hundred dollars for the Turtle Patrol. She just thought it was a wonderful thing."

The following statement of gratitude is posted at the end of the Patrol's "About" page on their website:

> *"We wish to thank everyone that contributed for their hard work and generosity. Your help not only allows us to keep sea turtles safe, but our volunteers as well."*

EXTRA! EXTRA!

With all of the conflicts and litigation surrounding turtle protection laws—coastal lighting and beach driving being the most prominent of these—sea turtles and, by association, the Turtle Patrol became no strangers to the press. Whether it's articles questioning the Patrol's methods and priorities or lauding them, in community

Cartoon featured in the *Daytona Beach News-Journal.*[19]

papers or city ones, turtle-related content can consistently be found all over the pages of the daily news.

In a lot of ways, this publicity has been an immense help in making the Turtle Patrol into who they are today, in passing legislation they've fought for and in electing new members. Headlines like the two below, for instance, have worked to effectively get the Turtle Patrol's name out there, informing the public of what they're doing and why they're doing it.

Turtle Volunteers Keep Watch Over Nests, Eggs[20]

NIGHT DUTY: On Turtle Patrol
Intrepid biologists and volunteers brave bugs and long, humid nights to protect sea turtle nests[21]

But in some cases, the attention they amass in the press can have adverse effects on what they're trying to accomplish. Take the beach driving controversy of 2004–2005, for example. In petitioning to make driving on Flagler County beaches illegal, compounded with the lawsuits of a turtle-loving local, Shirley Reynolds, so much noise was made of the issue that it caused local officials to tire of the subtleties and do what they thought would most quickly and cleanly shut the entire drama down: they made beach driving illegal, to everyone—even the Turtle Patrol.

From that point forward, the Patrol was told they would have to forgo their ATVs for their boots while surveying. Sure, this solved half of their problem—now residents were no longer allowed to cruise the shoreline, spilling fluids into the surf, compacting the sand, running over marked nests or crushing hatchlings and thumb-sized baby birds. But now, on foot, carrying their stakes and ribbon and hammers and paperwork and coolers, it was more or less impossible to monitor the entire twelve-mile length of Flagler coastline for which they were responsible. Some beaches didn't even have access points or were behind gated communities and private properties. In these areas the beaches couldn't be reached at all. They were simply lost.

In *VHI Behind the Music* terms, this was the Turtle Patrol's downward spiral, their "dark days." Survey areas were no longer being fully monitored; nesting statistics were not being recorded, leaving gaps in the records to this day. There were even run-ins with the law. Some Patrollers stored supplies in friends' home garages along the beach so they wouldn't have to carry them all during their miles-long monitoring walks; others hid what they could in dunes and under walkovers, scrounging in the brush for retrieval later. Things came to a head when one day in 2006 the Patrol got word of a stranded green turtle miles down the surf and, knowing that there would be no way they would be able find it in time on foot, decided to flaunt their new regulations. In a surge of animal activist-inspired civil disobedience, Lori Ottlein and another Patroller, Denis Cody, had had enough and, with the blessings of their principal permit holder, set out on their ATV to Varn Park to save the stranded animal. In no time, the police arrived and promptly booked the couple, bringing them "downtown" to face the local tribunal.

Lori Ottlein recounts the arrest: "We were just like, this is ridiculous. So we took the ATVs and there was supposed to be a green that was in distress, up there in Beverley Beach. We said, 'Well, we're just gonna do it, because we have to go save her.' . . . 200 pound turtle—what are we gonna do? So we just did it. And that was when we got arrested."

Actually, it was Denis who was arrested. The police did remove the two and their vehicle from the beach, but because Denis was the one driving, he was the one handcuffed and tried and turned into a martyr. "And they were gonna put him in jail!" Ottlein remembers. "*That* was amazing."*

At this time, Florida Fish and Wildlife Conservation Commission (FWC) biologist and turtle expert Dr. Robbin Trindell (the woman who issues the Patrol its permits) traveled from Tallahassee, Florida, to argue on the Turtle Patrol's behalf. She claimed that if the county was disallowing their driving of ATVs on the beach for Patrolling, they were, in essence, disallowing the conservation of an endangered species, an affront to federal regulation. This lead to a series of Patrol protests on the steps—or, lawn—of Flagler County's Court House, the first of which, the one Ottlein refers to as "the big protest," taking place the day of Denis Cody's trial.

But as the underdog, or the celebrity (whichever you prefer), garnering sympathy is usually a fairly easy task, and after about three months of hearings and newspaper involvement, Cody's charges were dropped. The Turtle Patrol's driving ban was also overturned, re-granting them their license to drive on the beach under the umbrella title of "emergency vehicles," after an entire season of lost time.

Just like that, the Patrol was back on top. But for some residents,

* Funny story: The morning of the arrest, Lori had arranged an introduction between Gail, a co-worker/friend of hers, and Denis because she thought they might make a compatible match. She had been talking Denis up for weeks, how great a guy he was, how much her friend would like him . . . And when Gail arrived at the beach for the meeting, she wasn't there an hour before the cops pulled up and hauled Denis away with his hands behind his back. Leaving after this worst first impression possible, Lori remembers her friend's words exactly: "No thanks, Lor," she said, then drove away. Later, Denis and Gail would meet again and, later still, be married. They are now living out of the country together in Panama.

their victory was an outrage, and understandably so to more empathetic members of the survey team.

"When they took the cars off the beach there were a few people who really complained," Danny Rutkowski remembers. They couldn't understand why the Patrol was allowed to drive but they weren't. "It's hard to explain," Danny says, "because they invested, a lot of these people . . . in four-wheelers just like we drive. And you know, a retired guy and his wife get on their four-wheeler, got to the beach and [would] go fishing. Say he's seventy years old, well now he has to park on the top and carry everything down. So it was an inconvenience for awhile. There were some who gave me a hard time."

Even today—a whole two, three years later—there are still those who make snide comments and voice their discontent. To them, driving their cars on the beach was a *right,* not a privilege, which is what they argued in court and county meetings while the decision was in progress. The newer or shyer surveyors cite few to no run-ins with embittered residents—the Ryans, for instance, say that most people they see on patrol know them by name and sometimes stop them for friendly conversation. The more extroverted ones, though, the ones known to hand out pamphlets on the beach or invite strangers to nest cleanings, feel the neighborhood backlash often. To those residents who made habits of beach driving, the act of taking it away was personal. It was something they had grown up with and had been doing long before the Turtle Patrol was around or sea turtles were listed as endangered. But now, all of a sudden, they couldn't do it anymore—yet Patrol ATVs were shuffling by every morning with absolute free reign. It didn't seem fair.

To Dale Hench the key is presenting a good public image in spite of antagonistic remarks or passive-aggressive behaviors from any disgruntled beachgoers, like the fishermen who cast their lines in front of his and his wife's path during morning surveys, or those who used to be friendly now looking away whenever they see a Patroller around town. "It's a privilege to be out on the beach," Hench said, "we can't abuse it. We're not out there to joy ride, and some people think we are."

It was the age-old story of a few individuals ruining the good time for everybody else. People would drive to the beach, many from out-of-town surrounding areas, and roar through the dunes, drive over nests, peel out and leave deep gashes all along the beach like pot holes. But this wasn't everybody. There were fishermen who minded their own business, carrying out routines they had in place for years, but because the county never entertained the idea of vehicle exceptions, now they were out of luck. And out of a routine.

Eventually, however, most of the hard time did pass and soon the Patrol was able to get back to business as usual with little disruption. In many cities, though, losing the ability to drive their cars on the beach was something residents had to get used to. For years, since the early 1900s, beach driving was sort of a coastal tradition, a tiny, corner-of-the-map chunk of Americana, representative of a certain freedom only possible to feel and find on the coast. Granted, the rows of vehicles lined down the sand might have made the beach look like a parking lot, but it was *their* parking lot. Even as far back as the Great Depression, relaxing in one's car on the beach was a free and easy escape from the pressures looming over the dunes and inside the city. Stock car races were even held on beaches in Volusia County until the Daytona International Speedway was built in 1959. This was something people always just *did*—and still do, in neighboring counties and on neighboring beaches. To take it away, for certain residents, was to take away part of the beach's mystique and allure, take away part of the *beach,* itself. Now, it wasn't just theirs as it had always been. Now, more than ever, it was clear: the beach belonged to the city. The people were just sharing it.

On the following page is an exposé published in a Daytona local paper just as the hammer was due to fall on that area's beach driving in early 2000. The column inspires a certain feeling of weight in the reader, a certain sense of history that is so soon to be lost. By looking back at pictures from each decade, every one different yet exactly the same, glimmering with the familiar shine of sunlight off chrome, it's nearly impossible not to "get," even if it's just partially, the attraction of keeping the beach as these people did for so many years. They were kings and queens out there, these people, each one laying claim to their very own piece of the world, ruling it, *owning* it, if only for just a couple of hours.

Interesting to note, too, is the width of the beach in these photographs, and how dramatically it has decreased over the years. Compare the beach's width in 1940 to that of only sixty years later, on the following page, in 2000.

12C — THE NEWS-JOURNAL **Tuesday, February 29, 2000**

THE END OF AN ERA

1920s

The World's Most Famous Beach as it was . . .

In 1902, a wealthy tourist staying at the Ormond Hotel noticed that bicycle tires left very small imprints on the hard-packed sands of local beaches.

He drove his Stanley Steamer out on the sand and began a tradition that has captured the imagination of motorists even to this day.

The one-mile stretch that will be declared off-limits to cars Wednesday to benefit a beachside development and a county parking garage has been one of the most desirable beach driving sites for millions of motorists over the years.

1920s

It has always been difficult to keep beach drivers in assigned traffic lanes, as the fascinating historical photo from the 1920s shows, above. Like the modern-day convertible, the open-seated roadsters of that era were popular 'beach buggies' of the Jazz Age.

And it wasn't just driving that made Daytona Beach famous, it was parking. The beach has always been a great safety valve for beachside neighborhoods, with its ability to park large numbers of vehicles.

1930s

As the popularity of the automibile increased, so did the notion of driving and parking on the beach. At lower left is a scene from a busy day during the 1930s. During the Depression, a carefree day at the beach was a popular family activity.

1940s

The Forties were festive on The World's Most Famous Beach. Holiday parades were among some of the inducements to bring family vacationers to the packed sands.

1970s

Many residents fell in love with this area while on Spring Break from college. The city used the area near the Boardwalk and Bandshell for concerts and other promotions. In the 1973 photo at lower right, Daytona Beach City Recreation Director Al Garguilo is at the wheel as two bikini-clad young women toss flying discs to eager collegians.

1900s

1930s

1940s

1970s

The End of an Era[22]

Last day to drive

Parking garage opens with fanfare, finance fears

By MATT GRIMISON
Staff Writer

DAYTONA BEACH — The nonprofit board overseeing the new beachside parking garage vowed Monday to conduct all its business in the open during its first meeting, which also featured ominous financial concerns.

The Volusia Redevelopment Parking Corp. board unanimously adopted a policy that it will follow Florida's Government in the Sunshine and public records laws even though technically it is not required, since it is a private entity.

Inside

Driving on Daytona Beach has been a tradition since the early 1900s. Historic photos document beach driving in the city's core area over the decades.
12-13C.

"It is arguable as to whether this corporation is subject to Florida public records laws," said Dick Kelton, who was elected as the corporation's president during the meeting. "It is certainly our intent to operate according to those laws."

The board heard some cautionary news about the expected financial state of the garage, which had its ceremonial ribbon-cutting Monday afternoon but opens officially Wednesday.

There is a $500,000 gap in revenue estimates from parking fees for the first year and possibly beyond. Management company Central Parking System Inc. has estimates saying about $1.5 million should come in, while the nonprofit entity set up to construct the garage — Community Development Properties-Volusia Inc. — figured the project would bring in about $2 million after taxes are paid.

If Central Parking is correct, the corporation would not be able to pay for anything but operating costs and bond payments, leaving rent to Volusia County and other bills unpaid.

"If we have a situation where we can only make debt service, we should be aware of that," said Dan Marsh, a representative of Community Development Properties-Volusia.

Volusia Redevelopment Parking Corp. was set up to take over ownership of the garage from Marsh's organization. That is expected to take place in the next several weeks.

Board members downplayed the financial danger, citing expectations that revenues will be more than Central Parking estimated when the garage and other aspects of a $200 million redevelopment in the core tourist district open.

News-Journal/Nigel Cook

Last day to drive.[23]

A headline from the same local paper, published just a couple of days prior to the exposé opposite, perfectly encapsulates, perhaps unintentionally, the bittersweet reaction that removing cars from the surf sparked in the town. The headline is paired with a cartoon of children burying a sedan underneath the sand, smiling as they push sand on top of it, their tiny bodies nearly standing in the shadow casting off of Beach Driving's graveyard headstone. Listen closely and you can almost hear the sand they're shoveling flurry down, crackling onto the hood like an early morning Florida sun shower.

Daytona's beach will never be the same.
It no longer can afford to be.[21]

What's especially fascinating in all of these tributes and retrospectives is that in other parts of the country, driving on the beach remains something residents never even consider. It was that way in the city in which Lori Ottlein grew up in New Jersey, for example, next to a beach she describes as eerily similar to that of Flagler County's. There, not only were vehicles not permitted but, similar to Florida's Crescent Beach, residents had to purchase "beach badges" before even gaining entry onto the sand. The beach of Ottlein's childhood was always pristine, as well, with city-sponsored clean-ups conducted every single morning before the gates were even opened. If beach driving isn't a *national* tradition, it begs the question of what made it a tradition at all, relegated to only certain quartered-off places on the map. What is it about the South, specifically, that opened the door for this kind of anomaly—the warmer weather, the longer summers, maybe? And is it precisely this singularity, this eccentricity with which the residents so fiercely identified? Was their struggle actually one away from the indistinguishable face of modernization, similar to the retraction of some away from the cultural whiteout of globalization? What makes Flagler FLAGLER, if not for its anomalies, these people must have wondered. If these changes occur, what will make our hometown different from any other?

In litigation is where the publicity attributed to the Turtle Patrol really has the greatest effect. Front page spreads of dead strandings elicit emotional responses and work to make boat and fishing regulations an issue that hits closer to home for people that either

Dead Turtle Washes Ashore[25]

didn't know a great deal about them or perhaps don't live on or near the beach. And more non-confrontational spreads, like *Turtles head back to beach: Dim lights—season begins today*[26] can be informational and helpful, as well. But reactions to these campaigns are not always positive, as was and is the case with the ongoing dispute over coastal lighting.

The issue of coastal lighting is one of those disputes that seems destined to never reach a definite ending, but instead a settlement and compromised. Because turtles get disoriented by onshore lights and sometimes follow them into dangerous locales, the Turtle Patrol and organizations like it have been rallying for years to have stricter regulations enforced on coastal residents and businesses. But many entrepreneurs fear that these limitations could potentially kill their business. Without attention-grabbing lights outside, who would know they were open, let alone want to shop or eat there? It's a debate many business owners simply refuse to back down from. For them, it's their livelihood that is at stake, not the simple matter of how bright or what color the fluorescents in front of their stores must be.

"I think (the Volusia Turtle Patrol) should put the shoe on the other foot and see what the business owner goes through," said one local oceanfront business owner, quoted in a 1999 edition of a local periodical. "I think they should have to give some sort of proof like pictures, before they can accuse a business." This quote is set directly next to a photograph of the individual standing stone-faced with hands in pockets, dead in the center of a well-lit commercial area along the coast. The article's title is juxtaposed directly above a much larger picture of a member of the Turtle Patrol studiously digging a nest. The header reads, *"Showdown looms over lights, sea turtles."*[27]

Finally, after more than ten years of fighting, and claims of countless sea turtle hatchling disorientations and deaths caused by the business owner's commercial lights, in 2009 the storefront in question became compliant with the county's lighting ordinances.

For years it's been like this, the arena floor shared between city, officials, and volunteers covered in patches and sheets of egg shells. After driving was banned and lighting ordinances passed in the county of Volusia, it became especially touchy right before hatching season was due to start. A headline from an Orlando paper maybe best captures the carefulness and near-hesitation with which "turtle people" and "regular people" at times relate: "Tensions hatch along with turtles."[28]

"Tensions—and a measure of distrust," the article reads, "stem from different ideas about how best to care for the federally protected species." But the conflict the story details feels, frankly, much more like an adolescent word-war than a truly "adult" discussion of opposing viewpoints. County officials, the article states, fear that "one of the volunteer groups counting nests in Volusia County" (A.K.A.: *the Turtle Patrol*) are juking their stats, using an unfair standard in determining what nests are "failed," simply to make the county look bad before state and federal agencies. In response, Turtle Patrol president Beth Libert turned the other cheek, praised the work of county inspectors for their enforcement of regulations in less commercialized areas, then alluded that the possible reason they might be dragging their feet in more developed regions—where bigger hotels are built along the Boardwalk—is because business owners tend to have higher levels of political clout than middle-class parents and school teachers.

As the article states, "Protecting sea turtles has been a longstanding source of controversy in Volusia."

But newspapers—the "action scenes" of Turtle Patrolling, the intrigue—can sometimes be misleading. It isn't all doom and gloom and mutual distrust among the locals—far from it. To regress, though newspaper publicity hasn't been enough to help the Patrol completely sweep the lighting issue, over time it has lead to a higher level of regard in the community. Many interviewed Patrollers did cite coastal lighting regulation as one of the major things they would change, government-wise, but they also gave examples of residents who shade or replace their outside bulbs accordingly every year, in full compliance. For the most part, many said, all you have to do is ask. If a nest is laid in front of someone's home, nine out of ten times they'll respect it and the Patrol's wishes—a respect that can only be attributed to the reputation the group has built over the years through their steadfast presence and tireless efforts.

Third-party publicity is only part of the collective voice the Turtle Patrol demonstrates in its community, however. The other major part comes directly from the literature they have available along the beaches they monitor, and for years, it used to come from their newsletter, the *Turtle Tracker,* which was published three times yearly.

An informative piece, the *Turtle Tracker* told of the largest issues facing the sea turtle community and what the Turtle Patrol was doing about them. Inside were success stories and sad ones, fundraising opportunities as well as volunteer activities, what to do if spotting a stranding and how to handle yourself around nesting turtles. The newsletter was all-inclusive, containing any and all information pertinent to turtles and, similarly, the Turtle Patrol. Wish-lists and honors were presented, and so were statistics, nesting trends and numbers for each new year the Patrol continues its surveys.

According to Turtle Patrol president Beth Libert, the Patrol's greatest success in all of its years of service is its nesting numbers. Reversing the downward nesting trend on the Florida coast from Volusia through Flagler County, she says, has been the most rewarding part of her job. And it's an achievement she shared with *Turtle Tracker* readers every newsletter. Following is a list of year-end totals from 1995 to 2009.

Nesting Numbers

North • *Volusia Turtle Patrol, Inc.*, covering the nesting area from Ponce Inlet north to Flagler County.

Nesting Survey Report	**1994**	**1995**
Total # of Nests	228	**211**
Total # of *False Crawls*	173	**112**
Total # of *in situ* Nests	147	**204**
Total # of relocated Nests	81	**7**

False Crawls are non-nesting adult crawls, crawls not resulting in a nest.

South • *Volusia Sea Turtle Society*, covering the nesting area from Ponce Inlet south to Canaveral National Seashore.

Nesting Survey Report	**1994**	**1995***
Total # of Nests	261	**218**
Total # of False Crawls	375	128
Total # of *in situ* Nests	199	213
Total # of relocated Nests	62	5

Nests left *in situ* are natural nest sites, nests left where they were laid.

*estimates, complete 1995 numbers were unavailable at press time

1995 Volusia County Stranding and Salvage Results as of September 1995

Caretta caretta *(Loggerhead)* - **40**
Chelonia mydas *(Green)* - **18**
Dermochelys coriacea *(Leatherback)* - **0**
Eretmochelys imbricata *(Hawksbill)* - **4**
Lepidochelys kempii *(Kemp's Ridley)* - **57**
Total 119

A stranding is a turtle that has washed up either dead or injured needing professional attention.

1995 State of Florida Stranding and Salvage Results as of September 1995

Caretta caretta *(Loggerhead)* - **397**
Chelonia mydas *(Green)* - **264**
Dermochelys coriacea *(Leatherback)* - **15**
Eretmochelys imbricata *(Hawksbill)* - **14**
Lepidochelys kempii *(Kemp's Ridley)* - **98**
Unidentified - **21**
Total - 809

Above numbers are from Florida Department of Environmental Protection, Florida STSSN Update, October 20, 1995.

Nesting Numbers

North • *Volusia Turtle Patrol*, covering the nesting area from Ponce Inlet north to Flagler County.

Nesting Survey Report	1994	1995	1996
Total # of Nests	228	211	261
Total # of False Crawls	173	111	122
Total # of *In Situ* Nests	147	204	240
Total # of Relocated Nests	81	7	21

False crawls are non-nesting adult crawls, crawls not resulting in a nest.

South • *Volusia Sea Turtle Society*, covering the nesting area from Ponce Inlet south to Canaveral National Seashore.

Nesting Survey Report	1994	1995	1996
Total # of Nests	267	232	239
Total # of False Crawls	375	319	341
Total # of *In Situ* Nests	199	224	230
Total # of Relocated Nests	62	8	9

Nests left *in situ* are natural nest sites, nests left where they were laid.

Figures are based on numbers reported to Florida Department of Environmental Protection.

1996 Volusia County Strandings and Washbacks Reported

Species	*Strandings*	*Washbacks*
Caretta caretta *(Loggerhead)*	76	20
Chelonia mydas *(Green)*	18	3
Dermochelys coriacea *(Leatherback)*	6	0
Eretmochelys imbricata *(Hawksbill)*	2	0
Lepidochelys kempii *(Kemp's Ridley)*	4	0
Totals	106	23

A *stranding* is a turtle that has washed up either dead or injured needing professional attention. A *washback* is a hatchling that made it to the ocean and then due to surf conditions was found washed back onto the sand.

These figures are those reports received by 12/31/96.

1996 Activities

Exhibits and presentations since May 1996.

May 26	Turtle Friendly Establishment, Sandpiper (NSB
July 1	Southeast Volusia Leadership Program
July 4	Turtle Talk, Museum Arts & Sciences
July 24	Volusia Vision Beach Committee
Aug. 1	Turtle Talk, Museum Arts & Sciences
Sep. 12	Women in Tourism
Sep. 21	Coordinator, Florida Coastal Cleanup
Sep. 23	Longstreet Elementary
Oct. 15	Interview, *Best Friends* Magazine
Nov. 9-10	Tails and Tots Pet Festival
Nov. 23-24	Celebrate Christmas in Old Daytona

If you would like to schedule a presentation, please call 767-5257 (north of the inlet) or 423-0083 (south of the inlet).

Nesting Numbers

- *False Crawls* are non-nesting adult crawls, crawls on the beach not resulting in a nest.
- Nests left *in situ* are natural nest sites, nests left where they were laid.

Visit our web site: http://www.volusia.org/tpatrol

South •***Volusia Sea Turtle Society,*** covering the nesting area from Ponce Inlet south to Canaveral National Seashore.

Nesting Survey Report	1994	1995	1996	1997
Total # of Nests	267	232	239	159 (155 Loggerhead, 2 Leatherback, 2 Green)
Total # of False Crawls	375	319	341	237
Total # of *In Situ* Nests	199	224	230	130
Total # of Relocated Nests	62	8	9	29

North •***Volusia Turtle Patrol,*** covering the nesting area from Ponce Inlet north to Flagler County.

Nesting Survey Report	1994	1995	1996	1997
Total # of Nests	228	211	261	187 (182 Loggerhead, 3 Leatherback, 2 Green)
Total # of False Crawls	173	111	122	108
Total # of *In Situ* Nests	147	204	240	158
Total # of Relocated Nests	81	7	21	29

Volusia County Nest Numbers

Sea Turtle Species	No. Peninsula State Rec. Area (2.7 miles)	Volusia Turtle Patrol (25 miles)	Volusia Sea Turtle Society (10.5 miles)	Canaveral Nat'l Seashore (11.25 miles)	Totals
Loggerhead	28	255	262	1470	2015
Green	6	6	9	147	168
Leatherback	0	2	2	3	7
Kemp's ridley	0	0	1	0	1
Total Nests	34	263	274	1620	2191

1999 Nest Numbers for Volusia County Monitoring Areas

Sea Turtle Species	No. Peninsula State Rec. Area (2.7 miles)	Volusia Turtle Patrol (25 miles)	Volusia Sea Turtle Society (10.5 miles)	Canaveral Nat'l Seashore (11.25 miles)	Totals
Loggerhead	36	347	279	1601	2263
Green	1	0	0	2	3
Leatherback	1	0	2	3	6
Total Nests	38	347	281	1606	2272

2000 Sea Turtle Nesting Totals

Sea Turtle Species	Gamble Rogers State Rec. Area (4miles)	Volusia Turtle Patrol (25 miles)	Volusia Sea Turtle Society (10.5 miles)	Canaveral Nat'l Seashore (11.25 miles)	Flagler County (18 miles)
Loggerhead	58	278	320	3892	318
Green	10	12	8	654	39
Leatherback	0	2	0	9	4
Total Nests	68	292	328	4555	361

Volusia & Flagler Nest Numbers 2001 –2009

**Loggerhead-CC, Greens-CM, Leatherbacks-DC*

Year	Flagler	Flagler Beach	Volusia
2001	179 CC 5 DC	77 CC 1 DC	242 CC 7 DC
2002	217 CC 33 CM	77 CC 8 CM	226 CC 13 CM 1 DC
2003	199 CC 6 CM 6 DC	112 CC 1 DC	188 CC 1 CM 5 DC
2004	118 CC 20 CM	78 CC	129 CC 7 CM
2005	134 CC 37 CM 2 DC	62 CC 8 CM 3 DC	129 CC 5 CM 1 DC
2006	146 CC 18 CM 1 DC	62 CC 2 CM	188 CC 11 CM 1 DC
2007	149 CC 73 CM	58 CC 14 CM	264 CC 18 CM 2 DC
2008	218 CC 26 CM	90 CC 2 CM	282 CC 18 CM
2009	172 CC 39 CM 6 DC	42 CC 3 CM 2 DC	163 CC 7 CM 3 DC

Beth Libert

President of the Volusia/Flagler Turtle Patrol:
1992–Present
Sea Turtle Lover & Activist: 1983–Present

"The Turtle Lady"

"Working to protect the sea turtles for twenty-six years has definitely given me a feeling of purpose. I can see the rewards of all the hard work."

"People often ask me why I keep doing this and what makes it worth it. Seeing the hatchlings crawl to the water is my reward."

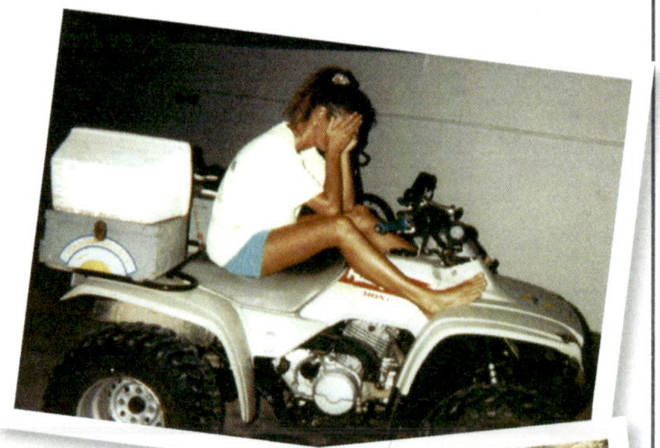

"My contributions to sea turtle conservation have made a huge impact on a large stretch of historic and critical sea turtle nesting beaches. Where there [were] only a handful of nesting females there are now hundreds and these beaches are much safer now [for] the hatchlings [to] reach the water. I have brought awareness and set precedents for change that not only protect our turtles but turtles on other nesting beaches throughout the state."

Trust

The sea produced an ancient form
with aquatic wings for soaring
that gouged the sand away from the tide
above the ocean's pouring.

She abandoned hope to trust the past,
heaved forth the future and at last,
buried it and left.

Now, two moons hence, little turtles pip,
with soft struggling bodies hatching.
The sands ensconce as eggs are ripped
by contorted masses scratching.

The siblings toil at a common chore
to whittle ceiling into floor,
until at sand's surface just short of sky,
the unsettled lie, becalmed.

The tangled turtles wait
as heat of day abates
and cool of night prods
their reluctance away.

At dusk the fits and starts begin
and through claw and strain,
above their heads sand rains again,
and yields to sky of night.

This army boiling in the night gains might,
and in waves, pours forth to see the sight.
Soft flippers patter and wipe sand from view
that eyes might seize upon the cue.

And then eyes do, they catch the glow
and every hatchling keen
rushes on to the goal they know
but they have never seen.

As if clockwork toys tightly wound
they keep their pace and bearing tight,
for unless the sea is quickly found,
they will not survive the night.

They choose their erring paths
with neither doubt nor anticipation,
and their consistency deals them life or death
with quiet resignation.

Thus, night wanes and sights of light
remaining
scatter throngs persistent
and about the dune abundant obstacles
restraining,
divide the dying from the spent.

Weakened few reach the sight they sought,
a deceptive brightness reassuring
where trusting forms are caught
by the sight of lights alluring.

Dawn now dries their searching eyes
and death now rests the weary.
Might fate have been more kind
to travelers more leery?

Were these turtles to awaken,
could they sense their mother's plight
having left her young forsaken
owing confidence in light?

Past's light offered not such bitter seas
nor played such deadly roles
to guide hatchlings on to sights like these
electric lights on poles.

Might we masters of the light adapt,
forgo complete control,
and lessen obsolescence
lest our presence take its toll?

To tread on earth with darkness soft
leaves not the night asunder
and preserves the stars and the moon aloft,
and obsoleted wonders.

—Blair E. Witherington, highly respected research biologist for the Florida Fish and Wildlife Conservation Commission at the Florida Marine Research Institute in Melbourne. His poem, "Trust," was published in the Fall-Winter edition of the *Turtle Tracker,* Volume 6, Issue 3/4

[another excerpt from the Hench interview]

Dale Hench: It's a great deal of personal satisfaction, holding a little turtle in your hand . . . We had a nest that was very close to the pier and it was the summer, a lot of tourists in town. And we're evaluating the nest, [which] happened to have a lot of little turtles in it. [So] we got the turtles in the ice chest—the same ice chest we might use to relocate a clutch—and then we release them all at the same time. And we had maybe not hundreds, but we had a lot of people standing around. And there was this one family in particular with a bunch of cute little kids, and they'd had a great time in Florida. They watched it, and they were taking pictures and taking notes. And then the mom and dad said, 'You know, we've done everything in Florida and the best time we had was—

Peach Hench: Including theme parks—

Dale: And the thing the kids liked the most was—watching this.

Peach: And they were going back home the next day.

Dale: Yeah, they were on their way home.

Peach: ' You've made our vacation.' . . . Just, you know, we've had that before. A lot of people on vacation say this is the best thing that happened on vacation.

They can never replicate that

Peach: No, never come close.

CHAPTER 6

Getting a Fix: Chasing the Turtle

The thrill of the chase: Patrollers on an ATV, looking for signs in the sunrise.

WHY WE DO WHAT WE DO

Like a still-born, like a miscarriage, any given nest being surveyed holds the possibility of never actually emerging. It might just lie there, underground as if already in burial. For months throughout the season, it will have seemed to incubate. Like tending to a pregnant mother, surveyors very well might have done everything right—protected it, nurtured it and kept it clean, unburdened—but come hatching time, the top table of sand may never shift. The beach, sectioned off and lumpy like sheets on a hospital bed, might lie dormant and cold under the sun. Patrollers did everything they could. Some people might hang their heads in these cases, sigh and chalk it up to fate. "It just wasn't meant to be," they would say. Others might break down.

All things considered, this pregnancy analogy holds nicely, what with the time and energy and passion put into every single egg inside every single nest on Turtle Patrolled beaches. On explaining this book project to a co-worker, this was the first idea that came to her. Nests are like pregnant women, she said. You watch them grow and grow and then give birth, and, in a way, their birth is partly yours, too. It's an analogy that stands maybe solely on the sentimentality it so flagrantly brandishes, likening beach surveyors to prospective fathers, pampering their wives and anxiously awaiting their first meetings with the new most important thing in each of their lives: their child. It offers a human and emotional edge to the relationship between Patrol and turtle, person and beach, an intimacy that almost seems requisite to understanding the "why" of everything these people do. The "where," "how," and "when"—those are easy, but "why"? It's a question that looms just behind every letter printed on these pages, behind every fact and quote and figure, nearly forcing the reader—or in this case the writers—to give greater credence to a sort of natural mysticism spoken of by no one in particular, not Patrollers, or those interviewed, or anyone.

Ask even the surveyors themselves and they'll admit: the Turtle Patrol, it's not *really* about the turtles. Most of the members just love nature and animals; others love the excuse to go outside, be with people, do something good. On the beach, Patrollers don't just watch over loggerheads and greens, they also pick up garbage and look after sea birds, supplying almost as much literature on them as on the turtles. Any animals found hurt or lost on the beach—weak and struggling raccoons, dead birds, dolphins, lost tortoises, baby crabs—the Patrol looks after them. "We are the eyes and ears," says Catherine Ryan, "we're [some] of the few people covering all the miles of [beach] every morning, from May until November. . . . Anything that gets washed up on the shore—we're the first people to see it, even if it doesn't directly relate to the turtles."

It's a connection they've forged with not one particular species but with the beach itself. And year-round they attach themselves to that connection, scaling its length in the summer months, smoothing out grains of sand on or around nests as if gently brushing the cheek of a sleeping infant, cleaning the land of washbacks in the winter, collecting genetics data, tasting the residual salt left on their lips hours after they've grown weary and headed for home. In these ways, the beach is always with them, specs of it constantly clinging on, in their cars, underneath their fingernails and in their hair, buried deep in the toes of their shoes, lining their heels, adding texture to their soles.

During the summer months, Patrollers eat together, wake up early together, watch the sunrise together. They plan small parties and convene for state-wide meetings, bring food, show off their new turtle knickknacks and accessories. These meetings are regulatory on paper, supposedly meant to update and organize, but speaking to individual members, they sound much more casual. They sound like something to look forward to.

Surprisingly, it's not just Turtle Patrollers that go to these permit holders meetings, either. It's also the "bird people," as Lori Ottlein calls them, and the "bear people," (although she jokingly admits, ". . . I don't really know why the bear people came—but that's okay!"). All of these different animal groups come together and they meet one another; they talk; they share stories. They go on and on, says Ottlein, talking about turtles and then breaking away a bit and getting different perspectives from other environmentalist groups.

When described by members of the Patrol, these meet-ups sound more like comic book conventions than business luncheons, more church gathering than work function, lazy Sunday mornings of good coffee and better company. They're college *a cappella* troupes and high school mathletes, a narrow cult convening *just* outside the fringe of "normal" society, amazed that in this huge world of shopping malls and button-downs they were somehow able to find each other.

"Everyone's like-minded and it's a good vibe," Lori Ottlein says of these congregations. "It just flows. It's like, somebody over here is not all, 'I hate the turtles.' Everybody's into it. No matter what you say, everyone's like, 'Yeah, yeah, yeah . . .' "

It makes it worth wondering whether in these talks, surveyors share with each other those tiny moments just before patrol begins, when they crest the dune for the very first time and initially glimpse the seascape, smell the air and wonder what adventures await them. It's worth wondering whether those seconds before their treads hit the beach, while ears still hear nothing but the soothing whisper of the sea, are communal or private, left as reflections for later, memories Patrollers stuff inside pockets and not dilute with words. Do they ever speak of looking to the fading night sky as they drive, closing their eyes and imagining themselves weightless, collapsing into grains and riding the breeze down to the surf like the earth flickering in a dusty trail from their tires?

Or do they keep it simple, nod, smile, and just go on pretending that what they're all really talking about, are turtles?

Conclusion

TURTLE POWER

What's important to remember is that, despite the Turtle Patrol's best efforts, if sea turtles cannot survive on their own, alone and without human aid, they are doomed. In this regard, what the Patrol is actually working toward is its own extinction and irrelevance, a sort of mirage in the distance they run toward with arms extended. The same as anyone looking after something larger than themselves—parents, Batman, or any other champion—the Turtle Patrol looks after and over its city with worrisome eyes, striving toward an odd and abstract goal, a goal many might not fully understand: simply, to not be needed anymore.

Patroller says goodbye to "Gus," a rehabilitated green she's releasing back into the harbor.

Maybe it's this similar faith in an unspoken abstraction that binds members of the Patrol so tightly together, that turns this group of mostly strangers into a uniformly tanned summertime family. Maybe it's this shared desire that makes this lifestyle into the "addiction" many describe, that drive them to hang leatherback icons around their necks like crosses, or burn loggerhead-shaped ink stains into their flesh, as if their bodies were no longer theirs, but instead the sea's, the spray's, the breeze's. Or maybe it's the habit. Maybe it's having a reason to get out of bed at 4:00 A.M. on a weekend. Whatever the motivation, there's a power in the water that these people cling to—that much is undeniable. And for some, it's a force they've been chasing for the better part of their entire adult lives, looking for it hidden two feet in the sand, waiting for it to show itself in the scurry of a hundred hatchlings drifting like a cloud into the water. Perhaps they're chasing an affirmation, a nod from nature acknowledging their fleeting presence within her.

"Why do you do it?" "How is it addicting?" Ask these questions to any given member of the Patrol and they're liable to shrug their shoulders and smile, stutter, ramble. "Why does anyone do anything?" they're probably wondering. "Because it's here. Because why not?"

Almost every Patroller's story starts differently but ends the same. It starts with seeing a volunteer on the beach and stopping them; or having a persistent friend; or loving turtles; or loving the beach. But it ends with body ink, holy icons, soft, glassy eyes and something that transcends rational expression.

Interestingly, of every Patroller we conversed with on the Web, a sizable portion held some variation of the exact same personal e-mail address—something with the word "beach," "turtle" or "patrol" incorporated into the user's "real world" first name (e.g. *PatrollerMike*). These were aliases, titles, chosen identifiers. And in a lot of ways, as tiny and unimportant as they looked on paper, they somehow seemed to better illustrate the mentality behind the Turtle Patrol way of life than any passage of elaborate prose ever could. To a large extent, these aliases represented what having a life's work must be about: *owning* a lifestyle, choosing something to love, and living for it. They reflected a pattern, a landscape on which many Patrollers have chosen to design the other facets of their lives.

Passions like these give the "addicts," as Lori Ottlein describes them, the "lifers," as Catherine Ryan coins, a calling, a wholly specific scope through which to see and recount their days. They're all-inclusive more-than hobbies that, like a colored filter, connect the sprinkled dots of time and experience together, making them all, every scattered moment and memory, radiate with one familiar hue that makes all the chaos of life and aging feel coherent. Cooking, writing, parenting, sports fanaticism, cinema obsession, bird watching, gardening, coin collecting, thrill seeking, work—these are all the same. They're a uniform and name tag. A chosen identifier.

It's an overarching theme, Turtle Patrolling, a mantra dusted in grains of gold. It's about the buff, the enthusiast being able to pick up any shell they find on the beach, resting its open end onto their ear, and hearing gusts of their own breath inside, their own gasps of excitement or despair, their own footsteps, pattering soft and fast somewhere in the distance, then gradually growing slower, closer and more labored, crunching down the sand and then stopping awhile to sit and ponder the waves.

Glossary

arribada: Translated from Spanish as "arrival by sea," an *arribada* is a large mass of either Kemp's or olive ridleys turtles rising from the sea to nest in synchrony on a crowded beach. No one knows exactly how or why this ritual occurs.

backstop: An approximated forty-five degree incline made in the sand as sand is pushed back with a turtle's rear flippers during the excavation of the primary body pit. Such a steeply inclined backstop is not present in the secondary body pit.

beach imprint: A map instinctually formed in a hatchling's memory during its first crawl from nest emergence to the water. This map brings female turtles back to their natal beaches to birth their own clutches. It is not scientifically proven how this map is made, but hypotheses suggest that the magnetic pulls of the earth play a large part in their creation.

beach nourishment: The process of scooping sand from a beach or the ocean floor and dumping it onto another beach to supplement the effects of erosion.

bekko: Japanese word for a turtle's shell-most commonly that of the hawksbill, but also associated with that of the green.

benthic grounds: Derived from the Greek word *benthos*, meaning depth, benthic grounds refer to anything on the bottom of a lake or sea, life associated with the ocean floor.

bleaching (of coral): The whitening of coral caused by a variety of natural variations, including: change in sea temperature, solar irradiance, subaerial exposure, inorganic nutrients, freshwater dilution, and several others. This whitening slowly kills off portions of a reef until the system is no longer inhabitable or self-sustaining.

bycatch: Any undesired extra catch brought up while fishing. Dolphins are often bycatch of tuna nets. Turtles are often bycatch of shrimp trawls.

carapace: The bony, hard top shell of a turtle, lined with rigid plate-like scutes.

Caribbean Conservation Corporation: An organization formed to ensure the survival of sea turtles within the Wider Caribbean basin and Atlantic Ocean through research, education, training, advocacy and protection of the natural habitats upon which they depend.

circle hooks: As opposed to "J hooks," circle hooks are a larger, more turtle-safe alternative that are said to result in less bycatch-related injury, death and entanglement.

Clean Water Act: Based on the Federal Water Pollution Control Amendments of 1972, the Clean Water Act (CWA), founded in 1977, regulates the amount of pollution in water supplies. Pollutants are classified as "priority" pollutants, including various toxins; "conventional," including biochemical oxygen demand (BOD), total suspended solids (TSS), fecal coliform bacteria, oil and grease, and pH (measure of acidity/alkalinity); and "nonconventional," including any pollutants not identified as either conventional or priority. The CWA regulates both direct and indirect discharges.

clutch: As dogs and pigs give birth to litters, turtles give birth to clutches. A clutch is the mass of eggs a female turtle will drop into a nest for incubation. They can range in size anywhere from fifty to one hundred fifty eggs.

coastal armoring: Techniques employed to maintain a beach's structure and volume after the effects of erosion. Sea walls or revetments are built to keep beaches from extending and coastal property from rust and erosion.

coastal lighting: Commercial or residential lighting set up along the coast, these lights are regulated to keep their glow away from beaches. If nesting female turtles and newborn hatchlings can see off-beach light, they can become disoriented, led away from the sea and toward inland.

coastal strands: A term used to describe an organization of flowering plant life that form along the shore in loose sand just above the tide line. When these plants are pulled out to sea, they serve as food and floatation devices for post-hatchlings making their first journey outward.

continental shelf: The undersea region of a continent that can extend several miles into the ocean. These areas are usually rich with marine life and resources.

convergence zone: Any area where two opposing forces meet and interact. In the ocean, this term refers to places where two opposite-facing currents converge, creating an ideal location for maturing turtles to float and feed, one where algae and other tiny organisms will constantly collect.

critical habitat: Any area that plays a critical part in a species development and sustenance, and is, in accordance, under higher governmental protection. For a species like the critically endangered North Atlantic right whale, for example, its critical habitat is established along the coast of southern Georgia and northern Florida, because that is where it births its calves. For sea turtles, these high-priority zones are established in separate vicinities around the globe according to the specific species in question.

dorsal: The top (back) portion of any animal (i.e. a dolphin's dorsal fin).

downwelling: A downward-facing flow of surface seawater brought on the wind-driven convergence of different water masses in the open ocean.

dredges: A process of excavation usually carried out, at least in part, in shallow near-shore waters. The goal of dredging is to transport sand and its nutrients to a different location.

dune: A wall, hill or ridge of sand on the beach, or in the desert, that forms naturally over time. Constantly shifting size and form with changing winds and tides, coastal dunes regulate a beach's construction and help to keep sand close to the water. Without dunes, erosion occurs more quickly, caushing beaches to lose volume more rapidly.

egg chamber: The cavity excavated by the rear flippers of a nesting turtle into which the turtle deposits a clutch of eggs.

emergence: The act of a hatchling breaking from its egg and emerging from its nest. Emergence usually occurs in groups, with one hatchling's climb to the beach triggering that of most of the rest of the clutch.

endangered: A species is labeled "endangered" when it is in great risk of extinction due to either dwindling population numbers or a change in environmental/predation parameters.

Endangered Species Act: Federal environmental regulation passed in 1973 designed to protect critically endangered species from extinction.

erosion: A phenomenon that occurs because of the effects of wind, salt, water, ice and gravity. These natural forces break down and move quantities of substances like rock and sand from one area and carry them to another.

escarpment: The perimeter of the secondary body pit where the front flippers have cut away a small cliff into the surrounding sand.

eutrophication: A spike in the concentration of nutrients in any given ecosystem environment to a degree that heightens the area's primary productivity. If this concentration is too great, it will adversely affect the environment, resulting in dramatic declines in water quality, fish and other animal populations.

false crawl: When a pregnant female turtle crawls onto a beach to nest but turns around and goes back into the water before dropping her clutch, she is said to have made a false crawl. These can occur because of human activity on a beach that a turtle perceives as threatening, or because of anything that unsettles her. If this process occurs too many times, and a turtle never feels comfortable breaching the shore to nest, she will drop her clutch in the ocean.

fibropapillomatosis: A herpes-like skin disease possibly caused by high toxin levels in the water. This condition is marked by benign yet debilitating tumors that can obstruct a turtle's movement, sight and breathing passages. There is currently no cure.

hatchling: Refers to a baby sea turtle newly emerged, breaking out of its shell and making its first crawl toward the surf.

gillnet: A common method sea harvesting device among commercial fishermen, gillnets are a type of net that is set in the water perpendicular to the ocean floor like a mesh wall. Fish attempting to swim through are caught in its holes and unable to free themselves. Turtles are common bycatch of gillnets.

Federal Water Pollution Control Act: The principal body of laws that gave rise to the Clean Water Act of 1977, a doctrine established to eliminate excess water pollution by 1985.

fibropapillomatosis: First discovered in 1983, fibropapillomatosis is a growing epidemic among sea turtles and thought to be in the vein of a herpes-type skin condition marked by benign but debilitating epidermal tumors.

flotsam: Refers to any loose trash or material floating on the water, possibly that from a ship wreck or its cargo.

Florida Fish & Wildlife Conservation Commission: An organization established to manage fish and wildlife resources for their long-term well-being and the benefit of people.

jetsam: Items deliberately cast off a vessel at sea in order to lighten its weight load or make it more stable in cases of emergency.

juvenile: In the context of this book, juvenile refers to young pre-adult turtles who are not yet of full size or sexual maturity. They are usually found drifting in the oceanic-zone post-emergence, or in nearer-to-shore coastal waters foraging and completing their growth process.

in situ: Eggs remaining in the nest in which they were birthed.

longline: Commercial fishing gear that hangs a dense collection of baited hooks underwater. These mechanisms can be miles long and are one of the leading causes of death and injury among loggerheads.

longshore drift: The perpetual movement of sand down a beach to replace it in areas where it has been eroded.

maritime hammocks: Occurring along the warm, southeastern tip of Florida, a non-coniferous forest with a canopy composed of a large variety of tropical tree species. These environments are essential to migrating birds, which rely on them for food and shelter after trans-oceanic and trans-Gulf flights.

National Marine Fisheries Service: A division of the Department of Commerce, this organization is responsible for the management, conservation and protection of living marine resources within the United States' Exclusive Economic Zone (water three to two hundred miles offshore).

neritic zone: Part of the ocean extending from the low-tide mark to the end of the continental shelf, the neritic zone refers to relatively shallow nearshore waters.

nesting: The process of a turtle returning to land and digging a pit in the sand in which to drop her eggs. Once dropped, eggs can incubate up to sixty days. The mother will never return to tend to her nest or eggs.

oceanic zone: Deep, open sea where hatchlings go after emergence to drift and forage away from near-shore predators.

pelagic: The pelagic oceanic zone refers to the area of open ocean lying over and beyond the continental shelf, while excluding near-shore and estuary areas. In describing a marine animal, pelagic refers to the free-roaming creatures that inhabit the water column above the seafloor but below the surface.

plastron: The bony plate that forms the underside of a turtle's shell. It is the orange-colored six-pack, in "Teenage Mutant Ninja Turtle" terms.

pots: Constructed, typically, of coated wire mesh, pots are fishing traps set individually on the ocean floor by line. They harvest various species of shell and finfish.

primary body pit: The excavation made by a turtle on the beach just prior to digging the egg chamber.

sargassum: Numerous species of brown seaweed distributed through temperate and tropical regions worldwide, where they generally populate in shallow waters and in coral reefs. This plant life is best known for its free-floating, pelagic state, when it is separated from reef systems and used as food and floatation for small, weak and growing species.

scutes: A layer of pliable protective plates on a turtle's shell that protect its torso while giving it a certain higher level of mobility than would a solid one-piece.

sea walls: Walls erected in place of an eroded dune to protect commercial or private property from the natural effects of erosion.

secondary body pit: An excavation made by a nesting turtle using the front flippers following the deposition of eggs. The spoil from the secondary body pit covers the primary body pit and the egg chamber with sand.

stranding: Any turtle that is washed onto the beach and unable to make its way back into the water and through the tide. Dead turtles washed ashore are known as strandings, as well.

threatened: Not every species cited in the Endangered Species Act is technically of "endangered" status. Those that are not, are labeled "threatened," which means they are vulnerable to extinction in the near future.

trawl: A device used in commercial fishing, a trawl is a large net that is dragged through the water to collect fish. Often used in shrimping, the use of trawls also result in sea turtle bycatch, which is why federal regulation mandates that only trawls with built-in Turtle Excluder Devices can now be used in the ocean.

tunicate: Commonly known as "sea squirts" or "sea pork," tunicates are sack-like filter feeders. They are usually found on the ocean floor but, as adults, will reside in the pelagic zone.

turtle excluder device (TED): Championed by conservationists looking for a solution to commercial fishing bycatch, the Turtle Excluder Device is a trap door built into a trawler net that allows turtles to escape before they are drowned or entangled.

upwelling: The wind-driven motion of cooler, denser and more nutrient-rich deep water upward to the surface of the ocean. This new supply replaces the warmer and more nutrient-barren waters at the surface, creating a hotspot for marine life as well as fishermen.

ventral: The bottom (underbelly) portion of any animal (i.e. the ventral side of a fish).

washback: Any turtle unable to fight tide and pulled ashore. Generally, washbacks are hatchlings who are pulled back to the beach by the waves while trying to make it out to deeper waters.

Turtle Sources

Caribbean Conservation Corporation: An Introduction to Sea Turtles, http://cccturtle.org/sea-turtle-information.php?page=overview

Florida Fish and Wildlife Conservation Commission: Fish and Wildlife Research, Institute, http://research.myfwc.com/features/view_article.asp?id=5182

Info on Green Sea Turtles (Hawaii), http://earthtrust.org/wlcurric/turtles.html

NOAA Fisheries: Office of Protected Resources, www.nmfs.noaa.gov/pr/species/turtles/

Sea World Sea Turtle Information book, www.seaworld.org/animal-info/info-books/sea-turtle/index.htm

See Florida Online: Florida Sea Turtle Information, www.seefloridaonline.com/turtles/

U.S. Fish & Wildlife Service: Archie Carr National Wildlife Refuge, www.fws.gov/archiecarr/photos/index.html

U.S. Fish & Wildlife Service: You Can Help Protect Sea Turtles, www.fws.gov/northflorida/SeaTurtles/20090700_You_Can_Help_ST.pdf

Volusia/Flager Turtle Patrol, www.turtlepatrol.com

Endnotes

1 *The Turtle Hospital: Helping Sea Turtles Survive.* "Hundreds of Sea Turtles Admitted." www.turtlehospital.org/blog/

2 *National Aquarium: Newsroom.* "Healthy Loggerhead Turtle Returned to Sea" http://news.aqua.org/2009/09/19/loggerhead-turtle/

3 PHOTO: Jensen, Michael P. "Olive Ridley Arribada in Mexico." NOAA Fisheries. www.nmfs.noaa.gov/pr/species/turtles/oliveridley.htm

4 PHOTO: Turtle Conservation Centre. "Olive Ridley Turtle." www.turtleconservationcentre.org/malaysian-turtles/olive-ridley-turtle/

5 NOAA Fisheries: Office of Protected Resources. "Marine Turtles." www.nmfs.noaa.gov/pr/species/turtles/

6 *Mote Marine Laboratory: E@News.* "January 2010: Rare Turtle Brought to Mote for Rehab . . ." www.mote.org/index.php?src=gendocs&ref=enews_January%202010&category=News

7 Pulver, Dinah Voyles. "Thieves pilfer sea turtle eggs in Flagler." *News-Journal,* July 31, 2008.

8 Duke University. "Duke Study Gives First Worldwide Measure of Sea Turtle Casualities by Longline Fishing." Science Daily (March 9, 2004), www.sciencedaily.com/releases/2004/03/040309072753.htm

9 *Ibid.*

10 Skoloff, Brian. "Groups sue federal government over sea turtles." *News-Journal,* April 15, 2009.

11 Evans, Ben. "Report: Loggerhead turtles at risk of extinction." *News-Journal,* September 3, 2009.

12 *Ibid.*

13 *Ibid.*

14 *The Graduate.* DVD. Directed by Mike Nichols. 1967: Embassy Pictures Corporation, 2005.

15 NOAA National Marine Fisheries Service. "Marine Debris Photo Album." www.pifsc.noaa.gov/cred/mdr_photo.php

16 Cuculiansky, Kelly. "Loggerhead turtles nest in record numbers." *News-Journal,* November 4, 2008.

17 Caribbean Conservation Corporation. "Threats to Sea Turtle Survival." www.cccturtle.org/sea-turtle-information.php?page=threats

18 *The Turtle Lady Legacy.* "Ila and Cousteau." http://theturtleladylegacy.org/1ila-Cousteau.html

19 Beattie. *News-Journal,* May 10, 1996.

20 Fernandez, Frank. "Turtle volunteers keep watch over nests, eggs." *News-Journal,* July 10, 2009.

21 Cuculiansky, Kelly. "Night Duty: on turtle patrol: Intrepid biologists and volunteers brave bugs and long, humid nights to protect sea turtle nests." *News-Journal,* August 31, 2008.

22 "The End of an Era: The World's Most Famous Beach as it was . . ." *News-Journal,* February 29, 2000.

23 Grimison, Matt. "Last day to drive: Parking garage opens with fanfare, finance fears." *News-Journal,* February 29, 2000.

24 Beattie. "Daytona's beach will never be the same. It no longer can afford to be." *News-Journal,* February 27, 2000.

25 "Dead turtle washes ashore." *News-Journal.*

26 Catron, Derek. "Turtles head back to beach: Dim lights—nesting season begins today." *The Orlando Sentinel,* May 1, 1999.

27 Call, Carie L. "Showdown looms over lights, sea turtles." *News-Journal,* August 22, 1999.

28 Catron, Derek. "Tensions hatch along with turtles." *The Orlando Sentinel,* July 15, 2000.

Index

U

W

About the Authors

Frank Gromling and Mike Cavaliere tag-team in this collaboration to create, what they hope, is a very different kind of nature book. Frank, a lifetime entrepreneur and nature lover, and Mike, an editor and creative writer, strove to make something special with this text—not just a book about sea turtles, but a piece that is as educational and coherent as it is eclectic and personal.

Please enjoy.